Charles West

On hospital organisation

With special reference to the organisation of hospitals for children

Charles West

On hospital organisation
With special reference to the organisation of hospitals for children

ISBN/EAN: 9783337161811

Printed in Europe, USA, Canada, Australia, Japan

Cover: Foto ©berggeist007 / pixelio.de

More available books at **www.hansebooks.com**

ON

HOSPITAL ORGANISATION,

WITH SPECIAL REFERENCE

TO THE

ORGANISATION OF HOSPITALS FOR CHILDREN.

BY

CHARLES WEST, M.D.,

Fellow, and late Senior Censor of the Royal College of Physicians of London;
Corresponding Member of the National Academy of Medicine of Paris;
President of the Royal Medical and Chirurgical Society of London;
President of the Obstetrical Society of London;
Founder of the Hospital for Sick Children, and for Twenty-three Years Physician
to the Hospital.

Published for the Benefit of the Hospital for Sick Children.

London:
MACMILLAN AND CO.
1877.

TO

THE RIGHT HONOURABLE

THE EARL OF SHAFTESBURY, K.G.,

PRESIDENT OF THE HOSPITAL FOR SICK CHILDREN,

THESE

SUGGESTIONS ON HOSPITAL ORGANISATION

ARE VERY RESPECTFULLY

𝔇𝔢𝔡𝔦𝔠𝔞𝔱𝔢𝔡

BY THE AUTHOR.

CONTENTS.

PART I.

GENERAL PRINCIPLES OF HOSPITAL MANAGEMENT.

	PAGE
INTRODUCTORY. . . . . : . .	1
THE COMMITTEE OF MANAGEMENT.	2
The Superior Officers	7
The Housekeeper.	8
The Superintendent of Nurses	10
The Director	11
THE MEDICAL COMMITTEE	13
THE HOUSE COMMITTEE	15
QUESTION OF SISTERHOODS	17
French Hospital Nurses	18
Sisterhoods in Catholic Countries Generally	25
Protestant Nursing Sisterhoods	27
English Nursing Sisterhoods	31

                                                    PAGE
QUESTION OF COMPARATIVE ADVANTAGE OF INTRUSTING

  NURSING TO SISTERS OR TO LAY NURSES 32

QUESTION OF INTRUSTING THE CONTROL OF HOSPITALS TO

  A SISTERHOOD 40

FINANCIAL RESULTS OF MANAGEMENT BY A SISTERHOOD . . 44

PART II.

DETAILS CONCERNING CHILDREN'S HOSPITALS.

AGE FOR ADMISSION OF PATIENTS 48

THE NURSES—

    Their Age 50

    Their Training 51

    The Religious Difficulty 56

    Their Relation to Head Nurse 58

    Their Number to Patients 63

    Provision of Extra Nurses 64

    Night Nursing 71

    Ward-Helpers 73

    Change to Convalescent Hospital . .   . 75

    Provident Fund 76

CONTENTS. ix

PAGE

THE DIETARY OF PATIENTS 77

THE WASHING AND LINENRY 79

THE OUT-PATIENT DEPARTMENT 83

CONCLUSION 95

ON HOSPITAL ORGANISATION,
ETC.

ON HOSPITAL ORGANISATION,
ETC.

PART I.

CAUSES which it does not concern me to detail here have now for a considerable time removed me from all share in the management of the Children's Hospital in Ormond Street, which I founded, and to which more than five-and-twenty years of my life have been devoted.

It has seemed to me that I might usefully employ the greater leisure which has thus been given me by putting together, for the help of those who may be privileged to carry on my unfinished work, a few observations on the Management of Hospitals in general, and of Children's Hospitals in particular.

In England, where all civil hospitals are independent of government control, and almost all derive their support from the charitable contributions of the public, the mode of management of these institutions must of necessity vary greatly, and can present nothing of that unity of plan which, counterbalanced though it is by various drawbacks, characterises most of the hospitals on the Continent.

The three large endowed hospitals of London are

B

under a practical dictatorship, with the great advantage that the dictator is resident; that for the most part he has nothing to do but to govern the institution over which he presides, and that the resources of which he disposes enable him to command the help of the most efficient subordinates. The days are long since past when the presidency of a hospital fell to the alderman who, when the position became vacant, chanced to be Lord Mayor of London; and when the treasurership was valued for the sake of the house attached to it; and for the privilege of being allowed to have a large balance of hospital money standing to the treasurer's own account at his bankers. In those days such institutions were managed by the steward, the clerk, and the different officials, with no intentional unkindness indeed to the poor; but with Talleyrand's motto, "*surtout pas de zèle*," as one principle of their conduct, and "charity begins at home" as the other; and a carriage and pair was sometimes maintained out of a nominal salary of a few hundreds a year.

In contrast to this, the treasurers of these great institutions are now earnest, intelligent, painstaking; each of them administers the funds as carefully as if he were Chancellor of the Exchequer with a clamorous Opposition to satisfy; and to these gentlemen are due many of the most important improvements that have of late years been introduced into hospital management.

Some of my remarks may possibly apply to matters of detail even in these larger institutions; but I am more familiar with the working of those other hospitals, the government of which is vested in a COMMITTEE chosen from the general body of subscribers.

The members of such a committee seldom bring to the discharge of their duty any special knowledge ; and as seldom feel the need of acquiring it. The Chairman is selected from among persons of rank or station, on account of the prestige which attaches to his name, or the weight of his character in the commercial world ; and in either case he is a real guarantee against the possibility of that malversation of which the annals of our charitable institutions some fifty years since presented but too many instances. He is chosen, however, but too often without any reference to the nature of his previous pursuits, and not seldom with the tacit understanding that he is not to be troubled about details. In consequence, if, as occasionally happens, the wheels of the charity get out of gear, and he is called upon as the *Deus ex machinâ* to set all right again, he finds himself confronted with questions concerning which he is profoundly ignorant ; and unable to exercise an independent judgment, he is but a puppet in the hands of any person both plausible and positive, or else the dupe of some philanthropical or sociological crotchet of his own which he has never tried by the test of practical experience.

The shopkeeper comes from behind his counter, or the man of business leaves his office, or the man of leisure saunters from his club and looks in at the meeting of the committee on his way home. He listens to the reading of the minutes, the list of new contributions, the number of patients admitted and discharged ; approves of a tender for meat or for coals, and sanctions the painting or whitewashing of the wards. Now and then he finds himself indeed intensely puzzled by some recommendation from the medical officers which entails an unlooked-for expen-

diture of money ; or by some complaint against the
management, or by some difference between the
officials.   His first feeling is one of displeasure at
whoever or whatever disturbs the placid dulness of
duties which otherwise would both rest his intellect
and soothe his conscience.   Soon, however, he succeeds
in shifting the burden of a decision from his own
shoulders, or resorts to that refuge of incapacity, a
compromise between the right course and the wrong
one, and then goes to his own house with a feeling of
honest satisfaction " that he is not as other men are,"
but that he regularly devotes a measure of his time,
as well as of his substance, to the service of the poor
and needy.

And all this is done in perfect good faith ; for it
never occurs to him that the members of a Hospital
Committee have other duties than can be performed in
the board-room ; and some indeed there are who, in
the course of twenty years, have scarcely wandered
beyond its precincts, content to take all on trust.   It
does not cross the mind of him who does this that he
has to do with a machine of extreme complexity, and
that to understand its management an apprenticeship
is as much needed as to understand the intricate
details of one of those pieces of mechanical contri-
vance which many of us gazed upon with vague
wonder a year or two ago at the South Kensington
Museum.

To this, which is no fancy sketch, there doubtless
are exceptions ; but it would be too much to expect
that busy men, and still more perhaps that idle men
should give up the time needed for a thorough understanding of the subject with which they are dealing.
And all deductions made, there is no doubt but that

committee men such as are described above do a certain amount of good. They prevent outrageous blunders, they interest others in the work in which they have interested themselves, and they do good to themselves by consecrating some small portion of their time neither to business nor to pleasure, but to the benefit of their fellows.

But while no blame can attach to the members of a committee for their individual ignorance of details which yet may be all important for the good management of a hospital, there yet is one respect in which censure is often most deserved. The members of these committees too often undervalue that knowledge which they do not possess, partly perhaps because they do not themselves possess it, partly because they do not see how perpetually it is called into play to meet this case or that, and partly because, feeling themselves—and justly—certain of the purity of their intentions, they have a firm conviction that to do good becomes one of the easiest, instead of being one of the hardest, things in the world, and that it implies something almost like a tempting of Providence to call to their aid the ordinary common sense which they never fail to exercise in their business or their profession.

The directors of one of those large hotels which now abound in London and other great cities, have no knowledge themselves of the details of its management; but they engage a manager whom they believe to be competent. They scrutinise his testimonials, they inquire into his previous history, they seek for proof of his past experience; and when appointed, if he knows his business, he follows the same course in the selection of his subordinates. The annual meeting of the shareholders, and the dividend declared, tell whether the choice has been well and wisely made, or whether the manager, in spite of the care with which he has been selected, has yet proved unequal to his responsibilities.

In the case of the hospital all these precautions are too often looked upon as needless; previous careful training is regarded as unnecessary, the wish to do good is taken to imply the power of doing it ; or experience gained in quite a different class of pursuits is supposed to be like a bill of exchange, convertible into solid gold at a few days' sight. And hence it is that so many hospitals and other charities, started under the best auspices, conducted with the purest motives, fall into decay, because those to whom its management is intrusted are unequal to the task. The annual meeting of the charity, when it appears that the cost per bed has risen by so much, or by so much, does not come home to the managers in the same way as the diminution of an annual dividend would do. He must be an exceptionally good man who would not be more vexed if he lost fifty pounds through the negligence of his servant than if the same amount of the funds of a hospital were wasted by the incompetence of a manager.

It should therefore be laid down as a rule subject to no exception,

THAT NO ONE WHATEVER SHALL BE APPOINTED TO A POST OF RESPONSIBILITY IN ANY HOSPITAL, FROM THE MANAGER DOWN TO THE HEAD NURSE IN A WARD, WITHOUT SATISFACTORY PROOF OF PREVIOUS TRAINING IN THE SAME OR IN SOME OTHER SIMILAR INSTITUTION, AND IN A CAPACITY THE SAME IN ITS NATURE WITH THAT OF THE OFFICE FOR WHICH HE OR SHE IS A CANDIDATE.

And, this principle laid down at the outset, we are now in a position to enter on the question—

To whom shall the Management of the Hospital be intrusted.

The moment we approach this subject, the complexity of the mechanism to be regulated becomes apparent. There are in every hospital three distinct departments—the medical, the economical, and the nursing ; each of which is to a great degree independent of the other, but unless all three act harmoniously together, the institution cannot prosper. Just as if a cogwheel has one tooth broken, it can no longer adapt itself to the other wheels, and the whole machine comes to a standstill.

The medical department stands most apart from the others, and to a great degree must be controlled independently by the medical officers of the institution, to whom the house-surgeons and the dispensers must be directly subordinate. But it touches closely the other two :—the consumption of wine, the expenditure on so-called dispensary and surgical stores, as spirits of wine, lint, oiled silk, bandage-calico, etc., constitutes a source of outlay in which the economy of the institution is concerned. In just the same manner an alteration in the diet table, or the too lavish ordering of extras of food, may swell the cost per bed in a way which, in the absence of any direct control, will put the hospital to heavy expense such as no subordinate can check, against which it will not be easy for him even to remonstrate ; and which the committee, necessarily ignorant of details, can only deplore and acquiesce in. Nor is this all ; but just as the medical officers may require a needlessly large number of nurses on some occasion, so on the other

hand the head of the nursing department may be inefficient, and may fail to provide competent nurses, or
may not have a staff in hand adequate to meet
probable emergencies ; or she may be disobliging, or
a mere *routinière*, who governs all by rules which are
inelastic, and which she does not know how to
modify so as to deal with exceptional cases.   In
such circumstances antagonism between the different
departments becomes chronic ; the head of one or
of the other gets the ear of the managing committee ;
the others lose interest in their work, and stand aloof
in dissatisfaction.   The machinery works indeed, but,
like a creaking wheel, there is much grumbling and
small progress.

It has been attempted to overcome the difficulty of
want of harmony between the economical and the
nursing department by making the former directly
accountable to the latter, and placing the HOUSE-
KEEPER or Matron under the immediate authority of
the Superintendent of Nurses.   But apart from the evil
of an indirect responsibility, the plan itself is not a
good one.   It would be difficult for it to work well
even if the superintendent of nurses were possessed,
as she very seldom is, of that kind of knowledge
which goes to make a good housekeeper in the wide
sense the term implies in the case of any one who is
to be the efficient head of the economy of a hospital.
But it generally happens that the superintendent has no
such knowledge (and it were unreasonable to expect
it of her), so that she is unable to correct the shortcomings of her subordinates, unable even to detect
them; hence an inefficient housekeeper may long retain
her post to the grievous waste of the hospital funds,
if she have but the tact to render herself personally

acceptable to the superintendent. On the other hand, an able housekeeper will soon detect her chief's ignorance; she will try to work independently of her; in doing so will find herself thwarted by the interference of a person who does not understand her business. She then either falls, for peace' sake, into the unwise plans of her superior officer, or bickerings arise between the two which end in the resignation or dismissal of a person who, had she been allowed to do her duty independently, would probably have discharged it extremely well.

The housekeeper is of course the inferior officer, and in an important sense is under the authority of the superintendent of nurses, just as an officer of the commissariat is under the authority of a general commanding a body of troops. She, with the co-operation of the steward in a large hospital,[1] or with the help of the secretary in a small one, has to provide whatever may be needed for the supply of provisions, linen, and all other non-medical requisites for the patients, and has the control of the laundry, the government of the servants as distinguished from the nurses, and has in short the conduct of the whole of the economical arrangements of the institution. She is not answerable for the too lavish requirements of the medical or nursing department, but she is answerable to the managers of the hospital for the honesty and intelligence with which those requirements are fulfilled. It is open to her to complain of the excess of the demands made upon her, or of

---

[1] A steward, or a man performing some of the functions of steward, is essential in every hospital for the control of the men-servants, and for many other purposes which it is unnecessary to enumerate. His relation, however, to the managers of the hospital is in all respects that of a male housekeeper.

difficulties thrown, either accidentally or ignorantly, by the other departments in the way of the performance of her duties; but these complaints should go not to the superintendent of nurses, who might suppress them, but to the managers, in other words to the committee, with whom must rest their investigation and their redress. Still further to insure her independent authority in her own department, her nomination as well as her appointment must rest with the committee, and must not be vested in any way in the superintendent of nurses.

The SUPERINTENDENT OF NURSES must be supreme in her own department, as the housekeeper in hers; bound to conform to the requirements of the medical department, or to produce reasons for not doing so; answerable, like the housekeeper, to the committee for her shortcomings, and, like her, having the power of remonstrance and complaint.

But her position is the most important one in the hospital; for that exists only for the cure of the sick, and their recovery depends on the efficiency of the nursing as much as on the skill of the doctor; often, indeed, good nursing is able to make up for deficient care or deficient skill on the part of the doctor. The tone, too, which she gives to the nurses pervades the whole hospital; and there is no position in the world in which the truly feminine virtues of humility, gentleness, simplicity, and lovingkindness have a wider scope, or yield a richer fruit. She also represents the hospital on many occasions to the public at large, and especially does she in the case of a hospital for children. Much correspondence of necessity passes through her hands; many of the supporters of the institution, for one reason or another, seek for an inter-

view with her, and are either won or estranged by
her manner of receiving them. She has it in her
power to quiet many complaints, often causeless
enough, on the part of subscribers, and to soothe the
anguish of relatives bereaved of those whom they held
most dear, or of parents deprived of their children.
She can be the good angel of the poor, the comforter
of the afflicted, the adviser, the friend of those who
work under her. It is hard to say what she may not
be and do if she will but content herself within the
sphere of her own duties : a sphere wide enough for
the head of a sage and the heart of a saint ; and if
she does but possess that first of all requisites for her
position—

> " A heart at leisure from itself,
> To soothe and sympathise."

Now if we lived in the golden age, if there were no
such thing as jealousy and strife, and foolish striving
after predominance, the heads of these three departments would be sure to work together harmoniously,
and no control nearer than that of the committee
would be required. But even in the noblest forms of
occupation there is risk of ignoble feelings coming
into play ; of indolence or incapacity discharging per-
functorily important duties, and, above all, of that
departmental jealousy which in small as well as in
large undertakings comes often into play, and mars
the work. Hence it is that there is now in almost all
hospitals, as everywhere on the Continent, an officer,
for the most part resident, who bears a kind of sway
over the others (under the name of house-governor,
DIRECTOR, clerk, or under some other designation),
whose duty it is to conciliate the reality of authority
over them with the freedom of action in most respects

of the heads of departments. He is a sort of surveyor-general, with the right and the duty to go everywhere, to see everything : not, indeed, to alter the regulations laid down by the committee, but to take care that those regulations are carried out in spirit as well as in letter. There is in the mind of many, and especially in the feminine mind, a strange impatience of control, a dislike to supervision, a wish to conceal from other eyes the way in which their duties are performed, and this not from a desire to shirk their performance, but from an intolerance of being told that those duties could be better discharged in a different way. Hence it is that institutions where no such control exists too often get into a state of hopeless muddle, while the committee, even when most intelligent and most painstaking, find themselves unable to detect the cause why the machinery, so carefully boxed in, works so unsatisfactorily. A director of some sort, then, is absolutely essential to the good working of a hospital the moment it has outgrown the dimensions of a home, and no one can visit the London Hospital in Whitechapel Road or the Manchester Infirmary without being struck with the impossibility of dispensing with the supervision of the house governor in the one case, or of the medical director in the other.

The exact nature of the functions of this officer and their limit must vary in different institutions, but they are sufficiently indicated in what has been already said. To those duties, however, as generally understood in this country it might be well to add that of drawing up an annual estimate or budget for the ensuing year, and submitting it to the committee at least three months beforehand, as the director of a

French hospital or *hospice* does to the *Administration des Hopitaux*. There is nothing that would better test the capacity of the government of a public institution than this looking *before* as well as after; nothing that would help the committee more to discover the weak places in the hospital administration, nothing that would keep all the officers better up to the zealous and intelligent performance of their duty, than the comparison at the end of a year of the actual results with the expectations entertained at its commencement.

In a French hospital the director is responsible to the *Conseil Général de l'Administration des Hopitaux, &c.* In England his responsibility must clearly be to the committee or governing body of the hospital. But the whole committee of management is a body too large and too fluctuating, and the members who attend its meetings are too changing to supervise the details of administration. And hence has arisen the necessity of two sub-committees: the medical committee for the consideration of purely medical questions; the house committee for the control of the internal management generally; both exercising a power of supervision and recommendation; neither having the power of laying down principles, or of originating any new mode of action, or of spending money.

The MEDICAL COMMITTEE should consist of all the honorary medical officers of the institution; for some only of their number can have a seat on the managing committee; otherwise they would exceed the number of lay members usually attending the meetings of the committee, and the management of the institution would be practically vested in the medical staff—a

result by no means desirable. The director should attend, though without a vote, the meetings both of this
and of the house committee. He should furnish his
list of agenda to the Secretary, who should add to it
any references from the managing committee, and
any question which a member might wish to submit
to the consideration of the committee.

The duties of the medical committee should include
the supervision of all the purely medical arrangements of the dispensary, and of the way in which the
dispenser performs his duties; and of the nursing
arrangements, in so far as their efficiency with reference to the patients is concerned. They should send
to the managing committee a quarterly report concerning the nursing in each ward, and concerning the
mode of performance of their duties by the individual
head-nurses, or ward-superintendents, and nurses,
coupled with any recommendation that may seem to
them desirable. In addition, no probationer should
be appointed a permanent nurse without the approval
of the medical committee, or without coming before
the committee to receive her appointment, which
should on every ground come from them, and not
from the superintendent of nurses, who is their subordinate.

In most hospitals where the most objectionable
plan of appointment to medical offices by the general
vote of the governors has been discarded, the nomination, which in the majority of cases is virtually the
appointment, rests with the medical committee.

To this, however, there are many objections; the
chief of which is that if a minority of the medical
committee are resolved on the support of a particular
candidate, it is almost always possible so to manage

the ballot, by the exclusion first of one and then of another, as to secure the return of their own favourite.

This risk and all complaint of unfairness, so fruitful a source of bickering, are completely avoided by the following plan:—A joint meeting of the managing committee and medical committee is held, at which the applications and testimonials of all candidates are read. The medical officers then, beginning with the juniors, each in turn express their opinion on the merits of the different candidates. If their opinion at all approaches unanimity, the candidate whom they recommend is sure to be appointed; while if they differ widely, the comparative merits of the candidates are debated among the lay members of the managing committee. In no case, however, do the medical officers take a part in this discussion, but, having expressed their opinion, retire, and leave the appointment entirely in the hands of the lay members of the managing committee. For many years this plan has worked with the best results at the Hospital for Sick Children.

But to return from this digression as to the mode of election of the medical officers. The duties of the HOUSE COMMITTEE bear the same relation to the general affairs of the hospital as those of the medical committee do to the medical department of the institution. According to the way in which it is constituted and the understanding which its members have of their duty, it is either a most important body or a mere *simulacrum*, a sham which stands between those who wish to manage the institution wisely and those who form its executive. If the latter seek for help, and frankly communicate their difficulties, and honestly ask for it in counsel with the house committee, that body is most useful. If, on the other hand, the house committee do not examine for themselves, do not insist on being made acquainted with everything, if they allow rules to be made without their cognisance, or suspended without their approval; if, in short, they do not form a distinct notion of their "*raison d'être*," they just shelter incapacity and perpetuate misrule, and never rise even when they do most above being harmlessly busy or foolishly fussy about details of no moment.

This committee should consist of one of the medical officers having charge of in-patients, and of two other lay members of the managing committee ; the former nominated annually by the medical, and the two latter by the managing, committee. The meeting should always be attended by the director without a vote, and by the secretary or assistant-secretary, to keep a record of the proceedings. The reports of the housekeeper and of the superintendent of nurses should be laid before each meeting, and each of them should be allowed to attend, and to express her opinion with reference to any matter in her own department whenever she desires to do so. The director should have the same power of representing any matter in any department of the hospital ; and should further have the right to submit any question concerning which either the medical or the house committee may have come to a decision opposed to his recommendation, to the further consideration and final decision of the managing committee.

Having thus passed in review the nature of the machinery by which a hospital must be controlled, we come next to inquire how its internal working may be best and most efficiently conducted, and what instruments are the fittest to carry on the work of charity. This brings us at once to the question, much debated at the present day:

Shall the nursing of the sick be placed in the hands of a sisterhood, or shall it be intrusted to lay agents?

The question is indeed by no means so simple as it would at first sight appear; for in practice it amounts in this country to little less than this—Shall the committee of a hospital retain the control of its internal working in their own hands, or shall they abdicate their authority, and intrust it to the self-devotion of ladies who, working without pay, are presumed to bring to the discharge of their duties higher motives and a greater conscientiousness than can be expected from those who fulfil their office as a means of gaining their daily bread? The clergy as a body throw the weight of their influence in support of the lady-workers, and say more or less distinctly that the care of the sick poor is not merely best discharged when religion pervades the spirit of the worker, but that the work itself is too high and holy to be performed save by those who have consecrated themselves specially to God, and for His sake to the service of the poor by a sort of female priesthood, who occupy by virtue of their work a higher level than that of their sisters in the world, or of those who profess to do the same work for their livelihood.

This opinion gains additional weight from the supposed identity of the claims put forth by Protestant Sisterhoods with those recognised in Roman Catholic countries as the rights of those Sisters of Charity who have made the name of St. Vincent de Paul a household word even in places where that of his Divine Master is scarcely known. But it is altogether a

C

mistake to suppose that the gentle sisters whom one sees in the hospitals of Paris, moving about the wards so silently, tending the sick so deftly, and with such unwearied care, are not only impelled by religion to the duties they themselves undertake, but are empowered to govern in its spirit, and to organise the whole service of the sick in accordance with their reading of its dictates.

This is not so; and it may perhaps help to a fairer judgment on the point at issue if we digress for a moment to examine a little in detail the regulations which govern the working of Hospital Sisterhoods in Catholic and Protestant countries respectively.

The whole administration of the hospitals of Paris is essentially secular. At the head is the *Conseil de Surveillance de l'Administration Générale de l'Assistance Publique*, with the *Préfet* of the Seine at its head. The active powers of this body are centred in certain administrators, at the head of whom with monarchical, almost autocratic powers is the so-called *Directeur de l'Administration;* while from the whole of the governing body, and from the whole of the administration, all clerical influence is, whether rightly or wrongly, carefully excluded.

At the head of each hospital is a lay director, or administrator, with one or more clerks, and in subordination to him a house steward, and whatever other such officers or servants are needed for the working of the hospital. He has, however, no direct control over the medical staff, though his representations to the *Administration Générale* on any subject connected with the performance of their duties would of course carry great weight.

The direction stands in immediate relation with the administration on the one hand, and with the head of whatever religious community undertakes the care of the sick on the other; and all complaints, either from the doctors on the one hand, or from the community on the other, would of necessity pass through him to the administration.

The arrangement between the community and the administration is terminable by a month's notice, and it is by no means unusual for it to be so terminated, while, since many communities devote themselves to the care of the sick, no grave practical inconvenience results from such changes. Thus, at Easter, 1874, the *Sœurs de Sainte Marthe* had just been removed from the charge of the sick in the *Hopital de la Charité*, and their place had been at once supplied by the *Sœurs de l'Hôtel Dieu*, as they are called from their Mother-House being within the walls of that hospital.

The sisters receive board, lodging, and certain articles of dress, and a right to support at the expense of the State in sickness or old age in one of the houses of the community; and the community receives 200 francs annually for each sister actually employed. For this they undertake the care of the sick, and certain duties connected with it, and usually, though not invariably, the superintendence of the kitchen and linenry. Their duties and rights are distinctly laid down in the *Règlement*, or convention entered into between the community and the administration. It is the same in principle in all the hospitals, but the details of that arrangement in different hospitals and with different communities naturally differ, and such details of course are not

made public.[1]   In the event of complaints arising
from any quarter with reference to any individual
sister, the Director would communicate with the
superior of the community, and the sister would be
replaced by another, in order to secure the good
working of the whole. Supposing—a most unlikely
supposition, that the head of the community did not
comply with the Director's wishes—the matter would
be brought by him before the Conseil de l'Administration for their decision.

In many of the hospitals, as already mentioned, a
portion of the domestic work, as supervision of the
kitchen, laundry, and garden, is performed in part by
the sisters, but in others those duties are discharged
by lay persons, and the sisters occupy simply the
position of ward superintendents. The under-nurses,
or *filles de service*, employed under the sisters are lay
women, who receive 20 to 25 francs a month, who do
domestic work, as scouring, etc., and who, though
employed also in nursing, are not of necessity educated
persons. The sisters have no power of appointing or
of removing the *filles de service*, who are appointed
and removed by the director. He, of course, would
listen to the complaints of any sister, and if those
complaints appeared to him not grave enough to call
for the woman's dismissal, would transfer her through
the means of the administration to some other institution, and thus provide for harmonious working
between the sister and her subordinates.

These *filles de service* are eligible for promotion,
first by increase of wages, and next by being transferred as *surveillantes*, head nurses, or superintendents

---

[1] The form of these conventions is given in full detail in the valuable
work of H. J. B. Davenne, *De l'Organisation et du Régime des Secours
Publics en France.* 8vo.   Paris, 1866, vol. i. p. 269.

to the *Maison Municipale de Santé*, or to the immense *Hospice of the Salpétrière*, where lay nurses only are employed. The reason for the employment of lay nurses in the two above-named institutions is that the former receives persons of very various creeds, who pay for their residence and medical treatment there, while the second is also partly an almshouse, partly a lunatic asylum.

In spite of the share taken by the sisters in the conduct of the nursing, and of the existence of a chapel in each hospital,[1] the absence of any religious element from the daily routine of the institution is remarkable. Mass is of course said daily in the chapel, and occasionally at altars in the wards of some hospitals, as for instance in the *Hopital Lariboisière*; but daily prayers are not offered in the wards, nor are the *filles de service* required to go to any religious service, nor even to be Roman Catholics, though as a matter of course the majority are Catholics; and a woman who habitually absented herself from mass on Sunday would not be regarded with favour by the sister under whom she served.

The priests attached to a hospital have no influence beyond what their devotion to their religious duties and their spiritual care of the sick may give them. The chaplain to the hospital is scarcely ever the

---

[1] No one was ever more alive to the wide distance which separates enforced religious observance from true piety than St. Vincent de Paul. Before his interference every patient admitted into the Hôtel Dieu was compelled to confess, and to receive Holy Communion; and Protestants, afraid of either being refused admission, or of being less well treated on account of their religion, confessed and communicated as well as the others. Vincent's first step in introducing the hospital visiting of his Dames de la Charité, catholic precursors of Mrs. Fry, was to suppress this abuse, and to substitute unobtrusive reality for pretence. See *St. Vincent de Paul, Sa Vie, etc.*, Par M. l'Abbé Maynard. Paris, 1860, vol. iii. p. 305.

spiritual director of the sisters, who are under the guidance of the director of their own community, and not infrequently there is even a little coldness between the sisters and the hospital chaplain.

In Paris, and through France generally, where the amount of intelligent lay nursing either in private or in public is but small, the general feeling both of the medical officers and the lay directors of hospitals is greatly in favour of the services of sisters as hospital nurses.

Limited, however, as is their sphere of action, and high as, beyond a doubt, are the motives by which the sisters are governed, the good is not without certain drawbacks ; if it were indeed, the workers and the work would be more than human.

The chief drawbacks alleged against their otherwise good service are—

1st. The occasional arbitrary change by the superioress of the community of one sister who had been in a hospital for some time for another less experienced, or to whom at any rate the ways and arrangement of the institution were strange.

2nd. The favouritism which was shown by the sisters to one patient rather than another, according as he was more or less attentive to his religious duties, and occasional ill-advised attempts at proselytism.[1] Even those who bring this charge, however, do not accuse the sisters of postponing the care of the body to that of the soul, and at any rate it is scarcely conceivable that in a children's hospital in England any Protestant sister would have so little faith in the

---

[1] See the cautions on this subject of M. Polinière, *Sur les Hopitaux, etc.* 8vo. Lyon, 1820, p. 38 ; and the same complaints re-echoed nearly fifty years later by Hippolyte Jaquemet, *Des Hopitaux et des Hospices, etc.* 8vo. Paris, 1866, p. 136.

Saviour's mission as to try to shake the simple belief of a little Catholic child in order to bring it to what she might regard as a purer creed.

3rd. The divided interest between the hospital, where their duties lie, and the Mother House, to which their sentiments and the greater opportunities of attending to their religious duties naturally draw them. The fact that the *Sœurs de l'Hôtel Dieu de Paris* have their Mother House in the hospital itself, so that the whole life and interest of the sisters from their novitiate is exclusively connected with the performance of hospital duties, has probably led to the general preference now given to them over the other orders of hospital sisters in the Paris hospitals.[1]

4th. And here it is perhaps best to quote the very words of a man who ranked high as a statesman, a philosopher, a philanthropist, and who was above all a devout Catholic, the late M. de Gerando.[2]

" Complaints are made in some hospitals that the sisters do not conform to the prescriptions of the medical men, whose functions they even usurp, and take upon themselves not only to modify the directions they have received, and to alter the diet, but even to administer medicines according to their own fancy.   Others ignore the rights of the civil adminis-

---

[1] These sisters, whose foundation dates back to the year 1217, are tertiaries of St. Augustin.   It is not without interest, in view of the claims to uncontrolled authority practically put forward by Protestant sisterhoods, to see how carefully any such predominance was guarded against by wise practical heads seven centuries ago.   It appears from the original statutes that " Two canons of the cathedral" (of *Nôtre Dame*, in the immediate vicinity of which the hospital is situated) " were appointed by the chapter with the title of *Proviseurs*" (or supervisors) "to maintain good order, while one of the thirty lay brothers was elected to take care of the hospital as superior, with the title of master, and he, with the supervisors, named a mistress for the sisters."
—*Histoire des Ordres Monastiques, etc.*   4to.   Paris, 1715, vol. iii. p. 186.
[2] *De la Bienfaisance Publique.*   8vo.   Paris, 1839, vol. iv. p. 345.

tration, place themselves even in open opposition to
the authorities, and excite some miserable strife
between the ecclesiastical service of the hospital and
the civil authority. Others again refuse to conform
to the regulations with reference to the accounts they
are desired to keep; or oppose the most manifest
improvements, as innovations which wound their
feelings; or with no ground whatever refuse to give
information, or to furnish explanations which would
conduce to the advantage of the work generally; or
on the other hand, allow themselves to be drawn
into intestine quarrels, or unreasonable exaggeration
of supposed slights or wrongs." He adds, with that
kindness which formed part of his nature, " These
reproaches doubtless are more seldom well founded
than some persons are inclined to believe," but still
he must have felt that there was no small measure
of truth in complaints which he states so fully.

A system of subordination of the sisters to lay
authority similar to that which exists in France
obtains also in all Roman Catholic countries on the
Continent. It will suffice to instance the hospital at
Munich.

This is under the direction of a physician appointed
by the King of Bavaria, with whom is associated an
inspector nominated by the municipality of Munich,
who is a sort of assessor, practical controller of
the institution, while the whole working of the
hospital is intrusted in all respects to the sisters of
St. Vincent de Paul, who act under the most minutely
detailed convention imaginable, in which are laid
down even to the number of the ingredients of the
daily diets of the patients, and who have no governing power, but merely authority to carry out· under

inspection from the municipality and control on the part of the government certain specified duties.[1]

If from the Continent we pass to Ireland, we find the same subordination to lay control in the Roman Catholic hospitals of that country. To this, indeed, there are two exceptions in Dublin, where the two hospitals, the Mater Misericordiæ and St. Vincent's, which were founded and are mainly supported, the former by the Irish Sisters of Charity, the latter by the Sisters of Mercy; are reasonably enough administered and governed by those sisterhoods without lay control. In the Jervis Street Hospital, however, although the nursing is in the hands of Sisters of Mercy, who have ward-maids under them for washing, cleaning, &c., the entire control of the place is in the hands of a lay committee of fifteen persons; and in the opinion of a very competent judge, himself a Roman Catholic, living in Dublin, the working of the Jervis Street Hospital is on the whole more satisfactory than that of the other institutions, which are entirely governed by sisterhoods.

From what has been already said it may fairly be regarded as an established fact that the position of the sisterhoods in Roman Catholic countries, so far as the care of the sick is concerned, is one purely of

---

[1] All these particulars are fully detailed by Thore, *Darstellung, etc., des Krankenhauses in München.*, 8vo., 1847.

It would occupy time, with no great advantage, to enter into further detail with reference to the hospital administration in Roman Catholic countries; for in all the subordination of the sisters to the administration is equally enforced. Those who care to pursue the subject will find information in Morichini, *Degl' Instituti di Publica Carita, etc., in Roma.* 8vo. Roma, 1835, concerning their management during the Papal Government, Verga, *Rendiconto dell' Ospitale Maggiore in Milano.* 4to. Milano, 1858; and the elaborate Report of M. Cerfberr, *Rapport à M. le Ministre de l'Intérieur sur les different Hopitaux, etc. dans les Etats de Sardaigne, de Lombardi, et de Venise, etc.* 4to., Paris, 1840.

administration within certain very narrowly defined limits, and under complete subordination to lay control, and that as a rule they possess no power of government whatever. The nursing of the sick has almost from time immemorial been so generally intrusted in Catholic countries to religious communities, that the means fail us for a minute comparison between their action and that of lay nurses in those countries in matters of detail. There can be no doubt indeed of the vast moral superiority of the sisters in all respects to the average lay attendants on the sick in most Continental Catholic countries; of which there can be no better proof than the fact that it is not yet forty years since the liability to corporal punishment for neglect of duty on the part of male attendants on the sick was still a regulation, though doubtless not acted on, in the hospitals of Vienna. The importance of the facts already adduced with reference to the position of the Roman Catholic nursing sisters in hospitals consists in this: that it shows how from their very foundation, and in countries where the whole population was in the completest sympathy with their creed and their practice, their functions were yet most rigidly defined and the limits beyond which they were forbidden to go were circumscribed with a care which some might fancy must have sprung from sectarian jealousy, rather than from the saintly piety and practical wisdom of men such as Vincent de Paul.[1]

---

[1] In Chapter iv. of the *Constitutions*, which he drew up in 1655 for the Filles de Charité, Vincent de Paul says, "They shall obey in all things not sinful, with submission of their own judgment and will, the bishops in the dioceses in which they are established. . . . the directors and doctors of the hospitals, in accordance with the rule of their order, as though at the first sound of the bell,—the voice of our Lord calling them."—Maynard, *Op. cit.* vol. iii. p. 226.

But we may now pass from Roman Catholic institutions to those which have been established in
imitation of them in Protestant countries.

The so-called Deaconess Institutions were established for the special care of the sick in imitation
of one of the supposed functions of the holy women
in the apostolic days of the Church. It is scarcely
necessary to tell the story of how the first of them
was founded in 1836 by a Lutheran minister, Pastor
Fliedner,[1] of Kaiserswerth, a man with much of the
piety and practical genius of Oberlin. It was connected with the Protestant Church as closely as
Catholic sisterhoods are bound up with the Catholic
Church, though in a different manner ; and this connection has been maintained with the Anglican clergy
by the many Protestant sisterhoods which have since
been established in this country more or less upon its
model, though with a gradually increasing tendency
to copy the exterior and to imitate the regulations of
Roman Catholic sisterhoods. The deaconesses are
equally bound by vows, though not for life, at least in
Germany ; and religious character as well as nursing
skill are essential for reception into their community.
Their machinery, in short, is of necessity in large
measure borrowed from that of Roman Catholic
sisterhoods, in spite of the antagonism which cannot
but exist between Lutheranism and Catholicism. In
its most fully developed form the community of
deaconesses was placed by the late King of Prussia,
Frederick William IV., under the oversight of an
order of female chivalry, the Order of the Swan, and
under obedience to the sovereign as absolute as that

---

[1] *Life of Pastor Fliedner*, by C. Winkworth. 12mo. London,
1867.

which in a Roman Catholic community would be yielded to the sovereign pontiff, and much more direct and immediate in its kind.

Under this influence the ideal of a Deaconess institution received its fullest development in the Bethany Hospital at Berlin.[1] Between it and any hospital in this country there is the important difference that the King is supreme, and governs the whole by means of a president of his own appointing. There is besides a managing committee of five men and three women, of whom the former were in the first instance selected by the King, the latter by the Queen. These hold their office for three years, and are eligible for reappointment. Besides these eight there are or were three permanent members, the chaplain, the lady superior, and Pastor Fliedner, who, however, died in 1864.

What changes, if any, have taken place since that date in the constitution of the Bethany Hospital are unknown to the writer; but the essential principles of these societies remain the same. The religious vocation is put first as constituting fitness for the work; while as the development of the individual both in technical skill as well as in Christian character is the object of the association, any hospital or institution with which a sister is connected is only a means to an end, and the mother house, not the temporary occupation, claims and obtains the full allegiance of each member.

It is obvious that with these principles the hospital is ancillary, a means to an end, and not the end for which the society exists; that, in other words, the

---

[1] The rules of the Society are given in full in Stein, *Das Krankenhaus der Diakonissen Anstalt Bethanien zu Berlin.* Folio. 1854.

improvement of the individual comes first, the care of the sick is subordinate to it.

In a little pamphlet published at Berlin in 1870 by Dr. Runge,[1] under the title " Shall the Care of our Sick be regarded as an Occupation for Lay Women, or as a work for Religious Communities ?" the whole subject is very fairly discussed, and four chief objections to the sisterhoods are distinctly brought forward —objections which, if they are of any value, apply far more widely than to the institution at Kaiserswerth, since they hold good with reference to the employment of Protestant sisterhoods generally in hospital work.

Dr. Runge alleges in support of his views :—

1st. The evil of a divided authority, the sisters owing obedience elsewhere than to the governing body of the institution.

2nd. The fact that the increase of personal sanctity, not the mere care of the sick, is of necessity the main object.

3rd. The fact—one might almost call it the consequence—that even at Kaiserswerth a *plus* of religion is allowed to make up for a *minus* of vocation or aptitude for nursing.

4th. The fact that since the death of Pastor Fliedner his successors at Kaiserswerth have developed the hierarchical character of the institution to the detriment of its other object.

He adds further the remark which the experience of all medical practitioners will confirm, that there is no necessary connection between religious feeling and the gentleness or fitness in other respects for the

---

[1] *Die Krankenpflege als Feld weiblicher Erwerbsthatigkeit gegenüber den religiösen Gemeinschaften*. Von Dr. F. Runge. 8vo. Berlin, 1870.

duties of a sick nurse, that the one may exist with, or, on the other hand, wholly independent of, the other.

Lastly, quoting from Dr. Varrentrapp, he sums up in his words the chief objections to the work of sister-hoods in hospitals.[1]

" 1. They isolate themselves too much from the other hospital officials, and occupy their minds with the idea of their own spiritual benefit to be gained by the active exercise of charity to their neighbours, rather than with the single thought of the good of their patients.

" 2. They are not sufficiently obedient to the doctors, but often place themselves in opposition to them.

" 3. For the same reason, from an unbecoming spirit of independence, with which is associated a certain spiritual pride, we meet much oftener in their case than among paid nurses with those who dabble in nostrums of their own."

His fourth objection, the postponing the bodily to the supposed spiritual good of their patients and the spirit of proselytism in which they too often indulge, happily scarcely concern us in the case of a hospital for children. They, however, are noteworthy from their almost complete identity with the faults alleged by Gerando. Catholic or Protestant, human nature and human frailties are the same ; and piety the most sincere does not render less true than in Chaucer's time the answer of the knight when called upon to

" Tell in audience
What thing that worldly women loven best.
My liege lady, generally, quod he,
Women desiren to have soverainetee."

_____

[1] _Lib. cit._ p. 17.

When, somewhat more than thirty years ago, religious zeal and charitable work acquired an earnestness and activity previously unknown in this country, the plan of the Deaconess institutions was introduced into England, and the help of deaconesses was had recourse to by many of the clergy to help them to work their parishes with greater efficiency.[1] Not much later other associations of women were formed with similar objects, but approaching more in their constitution to that of Roman Catholic sisterhoods, whose dress they imitated, and many of whose observances they adopted. The two kinds of associations corresponded in great measure with the requirements of the two great parties into which the Anglican Church is becoming each year more and more markedly divided ; the deaconesses carrying out the views of the so-called Low, and in a measure also of the Broad Church ; while the others, such as those founded by Miss Sellon at Plymouth, or now existing at East Grinstead, or in connection with All Saints' Church in London, conforming rather to the opinions and practices of what are termed the High Church and Ritualistic clergy.

It is no part of our purpose to follow the history of the good accomplished by means of all of these associations among the poor, the suffering, and the ignorant, working, as they have done, hand in hand with the clergy. The sisters of the latter class appear to have taken the more prominent share in the work of nursing the sick in hospitals, and it is with this only that we are concerned.

Two distinct questions here present themselves for our consideration :—

[1] *Deaconesses, etc.* By J. S. Howson, D.D. 18mo. London, 1862.

1st. Are the sisters on the whole *better* hospital nurses than those who undertake to discharge the duties of nurses as a means of gaining their living?

2nd. If so, is it desirable that they should occupy a position different from that which is assigned to them in other countries, and have the whole or almost the whole of the practical control over the institutions to which they devote their services?

The two questions are quite distinct. Some will negative and others affirm both, while many who answer the first in the affirmative may yet give a negative reply to the second.

The first question then is—

Are the sisters *better* hospital nurses than those who undertake the same duties as a means of gaining their living?

Now it must always be borne in mind that the position of England in this respect differs from that of the Continental countries, and especially of those Continental countries whence the model has been derived on which, to a considerable extent, the Anglican nursing sisterhoods have been framed. In them almost all hospital work has been done for centuries by sisterhoods, very rarely and in exceptional circumstances by lay agents.[1] In England, however,

---

A singular exception to the general rule is furnished by the hospitals of Lyons, where, in spite of their quaint nun-like dress, the sisters as well as the male attendants belong to no religious order, are bound by no vows, and have it in their power to return to private life, and even to marry.

Some of the remarks on this subject by the able historian of the large hospital at Lyons, Dr. Pointe, bear so closely on the questions under discussion that those who read it will feel no apology to be necessary for the length of this note.

"Our *Sisters* and our *Brothers*," says the writer, "not coming from any professed house whose rule they have to follow, and where they are sure of finding an asylum, or a refuge in case of need, depend more entirely

until within the last quarter of a century almost all
hospital as well as all private nursing has been done by
persons who followed the occupation of nurses as a
means of gaining their living, and these people have
shared to a great degree in the general increase of
education and intelligence which has pervaded all
classes of society. The general substitution for them

on the administration, and lend themselves more readily to the requirements of the service. The administrators exercise a more individual
influence over them ; their control is more direct, and consequently it is
easier for them than it otherwise would be to modify or change those
parts of the service which may seem to them to require amendment.

" It is evident, then, that this so to say only semi-religious character of
the sisters and brothers has the unspeakable advantage of placing the
immediate workers more completely under the hand of the administration, which can thus all the more readily keep alive in them that activity,
as well as that spirit of charity, which are both so necessary for the
performance of their duties.

" The servants " (in other words the sisters and brothers) " of both
sexes are taken from the families of artisans of known respectability,
who have been able to give their children an elementary education.
These persons then come to the Hôtel Dieu with habits of labour, and
with education enough to be placed in employments calling for a certain
degree of intelligence and knowledge ; or in default of their capability
for this in some cases, then at any rate in positions requiring physical
strength.

" I have stated that the *servants* often change their duties, and the
principle on which this change rests is a positive good. By means of
these frequent changes, all become soon *au fait* with most of the
branches of the service, and thus come to look on none as beneath
them ; and in the same way, too, petty jealousies and rivalries are pre-
vented from arising among them with reference to the different duties
of each, since they are shared indifferently and successively by all.

" It is especially in the case of the superintendent sisters, or head-
nurses, *sœurs cheftaines*, that one realises the full utility and importance
of this mode of organisation. In a word, a *sister* intrusted with the
entire management of a ward acquits herself better of this duty for
having herself gone through all the details of the work which she is now
called on to superintend. Thus, for instance, the administration of
remedies is managed all the better if the *sœur cheftaine* has herself been
engaged for several years in their preparation in the dispensary.

" Lastly, these sisters entail on the institution to a far less degree than
is the case with members of a religious order the inconvenience of
establishing themselves in the house as fixtures, whence they can never
be removed, even though they should fail in the proper discharge of
their duties."—*Histoire du Grand Hôtel-Dieu de Lyon :* par J. P. Pointe.
Large 8vo., Lyon, 1842, p. 106—108.

D

of gratuitous workers would have the result of depriving a large number of women of the means of earning their bread. Before such a proceeding is attempted, it should be shown first that an adequate supply of gratuitous workers can be depended on, and next that paid nurses are as a class not only much inferior to those who work without charge, but also that they are incapable of improvement. M. Jules Simon,[1] in his little book on the workwoman in France, dwells, and in no illiberal spirit, on the injury inflicted on the needlewoman and dressmaker by the competition they have to undergo with the convents and the work rooms, or *ouvroirs* founded in connection with them. " The competition," he says, "perfectly loyal though it be, and founded on the principle of association, on the very principle of liberty, is none the less crushing in its operation on those classes."

But then in justification of this competition, over which all who have mixed much with the classes whence the better order of sick nurses come must have heard many poor women lament, there comes into play the sentiment—may it not be called the false sentiment ? —connected with the office of a nurse. Her functions are exalted far above those of the doctor ; and the very people who when ill themselves would send for the most skilful medical man within their reach, irrespective of his religious tenets, or of whether he had any religion at all, speak and act, and this in perfect good faith, as if the first qualification of a nurse in a hospital were that she should be almost a saint. They talk as if there were something in the very touch of money that of necessity debases an act, and renders it impossible to serve God in one's daily task ; as

---

[1] *L'Ouvrière*, par Jules Simon, 18mo. 8th ed. Paris, 1876, p. 271.

though St. Paul had not been supported by the churches which he founded ; as though holy women had not ministered to our Lord.

And next we are told of the refinement of the lady, as contrasted with the coarseness of the nurse. But true refinement is something else than varnish, or than outward grace ; it comes from within, it is the appanage of the "pure in heart," the stamp of true gentility; it may be found in the poor seamstress or in the nurse as well as in the high-bred lady; it is independent of all conventionalities, is part of the habit of the mind, and

> " When a soul is found sincerely so,
> A thousand liveried angels lackey her,
> Driving far off each thing of sin and guilt."

Add to this refinement of the mind the graces of the accomplished gentlewoman, and one has all that is most beautiful, most perfect. Can these qualities, however, find no wider scope elsewhere, are they needed for the work to be done in a hospital ?—and if we could enlist them exclusively in the service, would it not be like harnessing the racehorse to the plough —like taxing Ariel to do labourers' work?

What did that wise man, St. Vincent de Paul, in such a case ?[1] He took the poor to wait upon and help the poor. A cowherd was his first *Fille de Charité*. He sought for poor girls who did not wish to marry, but who had no dowry to enable them to enter a convent, and were willing to devote themselves to the care of the sick poor. And from these small beginnings rose that society which now extends through every land, of lowly people doing lowly

[1] Maynard, *op. cit.* vol. iii. book vi. chap. i. pp. 185. 297.

work, of the poor helping the poor, and all for His sake, who though He was rich, yet for our sakes became poor.

And those who best know the poor will probably say teach, train, elevate the nurse as best you can ; but still the truest sympathy, the most practised skill in nursing the poor will be found among the poor themselves. Make the occupation an honourable one; take from it anything which would deter the gentlewoman from bearing part in it if driven by necessity to earn her bread, or if impelled by love of others to take a share for a time in humble duties ; but do not foster the delusion that those are best fitted for the work or will do most good in it who find or fancy that there is no scope for "doing their duty in that state of life to which it hath pleased God to call them ;" who think they serve Him best by leaving the position He placed them in ; and trading, not with the talent with which He intrusted them, but with some other for which it took their fancy to exchange it.[1]

[1] A wide divergence on many subjects from the opinions and teachings of the late lamented Charles Kingsley will yet not prevent most persons from recognising the wisdom from every point of view of the advice he gave to an unknown correspondent who consulted him with reference to entering a Protestant sisterhood.

"MADAM,—Whenever we leave the station where God has placed us, be it for never so seemingly self-sacrificing and chivalrous and saintly an end, we are yielding most utterly to that very self-will which we are pretending to abjure.   As long as you have a parent, a sister, a servant, to whom you can do good in those simple every-day relations and duties of life, which are most divine because they are most human, so long will the entering a cloister be tempting the Lord your God. . . . . But, Madam, be sure that he who is not faithful in a little will never be fit to be ruler over much.   He who cannot rule his own household will never (as St. Paul says) rule the Church of God ; and he who cannot keep his temper, or be self-sacrificing, cheerful, tender, attentive at home, will never be of any *real* and permanent use to God's poor abroad.

"Wherefore, Madam, if as you say, you feel what St. Francis de Sales calls a 'dryness of soul' about good works and charity, consider

It may be said, however, the purport of these observations is to exclude the highly-educated from the care of the sick in hospitals, to exclude them from a field in which their very education would enable them to exercise a far wider moral influence than can possibly be possessed by the best skilled ordinary nurses. But this is by no means intended. If, having first made sure that no nearer duty is left undone, they choose to undertake lowly duties, and to occupy a lowly position because they feel that in it they can best serve God, or because they wish to fit themselves by a sort of apprenticeship for higher duties, they will do nothing but good. But to do it they must pass through the same training as any other nurse, in as much detail and for as long a time. They must first learn to obey before they are fit to command, and their selection for the headship of a ward must depend entirely on the aptitude they display; while education and special aptitude by no means universally go together. To receive ladies into a hospital with no previous knowledge, and to train them, as is done in some, for two or three months to be *heads of wards* is a mistake in principle which insures maladministration, and confers power and responsibility on the incompetent. In a hospital there should be no governing caste; all officers should rise from the ranks.

"The sisters," says Miss Nightingale,[1] "must not be the heads of wards merely in order to use 'moral influence,' as the inexperienced sometimes fancy will

well within yourself whether the simple reason (and no shame on you !) be not because God does not wish you just yet to labour among the poor, because He has not yet finished educating you for that good work, and therefore will not let you handle tools before you know how to use them."—*Charles Kingsley : His Letters, &c.* 8vo. London, 1877, vol. i. p. 341.

[1] *Notes on Hospitals*, 3rd edition. 8vo. London, 1863, p. 183.

be sufficient. If a lady has in addition the same knowledge and experience as an old-fashioned hospital head nurse, which indeed many nurses, but only in secularly governed hospitals, have, good ; she is fit to be sister or head nurse ; if not, not."

A lady of high education, and of easy fortune, who comes to train as a nurse in a hospital, will find that she has much to unlearn as well as to learn. The recollections of the nursery where she grew up, the luxuries, modest though they may have been, which surrounded her childhood have to be very much forgotten if her new duties are to be well and wisely done. Simplicity in all things is the motto which should never for a moment be lost sight of, and this all the more if children are the patients with whom she has to do. It is not only that the expenditure of the hospital is often vastly increased by a departure from this principle, and by the small needless extras of diet, each of which, insignificant in itself, swells to a large total in the year ; but the effect of the over-indulgence is most injurious to the child itself. It is so easy to spoil a child, the process in the doing is so pleasant to both parties, especially when the spoiler does not see the full result of the mischief done. But when a little child, indulged, fondled, treated as if it were a prince, cries at returning home again, one cannot but ask oneself whether it were well done to make a hospital pet, and to send the little one back again, cured indeed of its bodily ailments by womanly tenderness, and care and love, but discontented with its home, unfitted to take its place again about the hearth where all had mourned its absence.

For a time the lady who has over hastily taken up with the new distraction, which she fancies a vocation,

may be satisfied with the new world of the hospital ;
with the affection long craved for, perhaps never given,
and which gushes up so abundantly from the child's
heart ; but then its sameness palls, and then comes the
desire to rule, the wish for power, the determination
to grasp it, the struggle to obtain it, based on the persuasion, not always a mistaken one, that those in whose
hands it is vested do not use it wisely. And then by
degrees the whole management of the institution
passes into the hands of those who possibly have had
no previous experience in the control even of so small
a machine as that of the household of which they
once formed a part.

The case, as thus stated, though a true, may yet
seem an extreme one ; but be that as it may, there is
no doubt but that a very large share of the internal
government of hospitals has of late years been
assumed by sisterhoods, or by associations resembling
sisterhoods, with no adequate provision, often with
no provision at all, for the adjustment of differences
between the governing body of those institutions and
the sisterhoods, which claim practically in many respects, a more than co-ordinate authority. It is not
many years since a most important institution was
shaken to its foundations by the endeavours to solve for
a time by action from without difficulties arising from
what was practically a contest for supreme power
between a nursing sisterhood and the lay governing
authority. A compromise between the two has been
arrived at, with the result common to most com-
promises of satisfying neither party, and of furnish-
ing no guarantee against the recurrence of similar
difficulties for the future.

It is evident then that the second inquiry proposed

is by no means an idle one, but that it imports us much to settle,

To what extent the practical control over a hospital should be intrusted to a sisterhood, or to an association resembling a sisterhood, or how far those who are engaged in the nursing of the sick should confine themselves to that alone, and should perform their duties in strict subordination to the general administration of the institution ?

The answer to this inquiry is by no means governed by that which different persons may return to the former. Some there may be who will prefer to intrust the nursing of a hospital to a sisterhood, either because they see no reason for discontinuing a practice which in this or that institution has already obtained, or because the proximity of the hospital with which they have to do to some established community facilitates their employment, while it furnishes an unanswerable guarantee for the high moral tone of the workers. Others, too, may estimate their influence so highly as to prefer them to lay-nurses, in spite of the drawbacks which have been mentioned above, as ward superintendents. On this account they may be disposed to underrate the greater difficulties of the control of a sisterhood, as of any other guild or corporation, owing to the solidarity between its different members, which will always lead one to screen the other, and the head to extenuate the deficiencies of his or her subordinates more than if they were simple individuals subject to a general rule. Or it may be that, weighing dispassionately the advantages and the drawbacks of the two systems, some may still arrive at the conclusion that the advantages of intrusting the supervision of the nursing to a sisterhood largely outweigh the evils which may be inseparable from it.

But an affirmative decision on the one point does not of necessity carry with it a similar determination on the other, and Miss Nightingale [1] while expressing her decided preference for intrusting the nursing to sisters, adds, and the italics are her own, "*provided the administration be secular.*" It must be added to this condition that the administration must be a *strong* one ; if it is weak, if, either from ignorance or from the *laissez faire* which creeps so insidiously into charitable institutions, *le roi regne, et ne gouverne pas*, the treasurer, chairman, secretary, or whoever may be the supposed head of the hospital, will soon be in the case of the old kings of France, with the head of the sisterhood as mayor of the palace.

The general administration of a hospital *cannot* be advantageously intrusted to a sisterhood, for if there were no other reason against it, the control of the nurses and the correspondence which that, of necessity, entails upon the superintendent of nurses, or Lady Superintendent, as she is properly enough called, is as much as one person can do well. But, besides, it must not be forgotten that in any religious community, the sanctification of its members is the main object ; the work of charity is but a means to this end, subordinate to, though largely contributing to it. How in the early ages of the Church the two objects were cultivated hand in hand, and with what beneficent results, need not here be told.[2]

---

[1] *Op. cit.* p. 183.

[2] The whole of the introduction to Montalembert's *Moines a' Occident* is perhaps the best and most eloquent defence of the monastic life of the middle ages to which any one can refer. Much too that is of interest will be found in Digby's *Mores Catholici*, book vii. of the 18mo edition

Those days, however, differed widely from ours; and at the time when St. Anselm was called from the abbey of Bec to take a share in the administration of a kingdom, nothing was more remarkable than the practical genius of those ecclesiastics who founded communities and organised charities; and whose regulations excite our astonishment by their astuteness as much as they themselves command our admiration by their sanctity.

Times have changed since then, and in nothing more than in the character of those who for the most part now enter these societies. The reader will judge for himself whether M. de Chateaubriand does not represent the present, as accurately at least as in his own eloquent words M. de Montalembert does the past.

"One of the strangest mistakes," says Montalembert,[1] "of many apologists of the monastic life has been to look upon it as an asylum for sorrowful souls, for those who are tired out, discontented with the world, incapable of holding their place where society has assigned it, eaten up with a sense of grievances or broken down with sorrow." "If there are places for the recovery of bodily health," says M. de Chateaubriand, "suffer at least that religion should have hers for the health of the soul, for that is far more liable to disease, and its sicknesses are much more painful, much longer, and much more difficult to cure." The idea is poetical and touching, but it is not true. The monasteries were never intended to receive the world's invalids. They were not the sick souls, but on the

of 1823; and many quaint touches of old customs and habits of thought with reference to the care of the sick poor in Monteil's *Histoire des Français des divers États*, edition of 1853, vol. i. p. 44; ii. p. 10; iv. p. 263; v. p. 286.

[1] *Op. cit.* vol. i. p. xxviii.

contrary, they were the healthiest and most vigorous souls that the world has ever produced, who presented themselves in crowds to people them. The religious life, so far from being the refuge of the weak, was, on the contrary, the arena of the strong. Who shall say that this is so now; or who, on the other hand, will deny that if there be any work from which all sentimentality should be most carefully excluded, any which calls for the robustest health of heart and mind, it is that of the organisation and control of a hospital ?

But, further : on the different sexes different gifts have been bestowed, that each may best supplement the other ; and daily experience does but confirm in each prosaic detail what, when we were young, we all loved to repeat in Schiller's graceful verse; and it implies no disparagement of womanhood, or of women, to state the fact that beyond the sacred circle of her home the power of organisation and control is not the one with which woman is most abundantly endowed. This power too seems to be one which its possessor does not owe in any large measure to mere culture, but, like the musical faculty, it must be to a great degree inborn. She who probably had more of this power than any woman in this country during the present century was the untaught Irish servant girl, Margaret Hallahan, whose name is venerated as Mother Margaret, the foundress in England of the third order of St. Dominic, and of all the Dominican convents in the land.[1]

It would seen invidious to go into particulars; to

---

[1] Any one whose sympathies are wide enough to find " good in everything " will read with interest and profit the *Life of Mother Margaret Mary Hallahan*. 8vo. 3rd ed. London, 1870.

instance the details of the working of one hospital in opposition to the results obtained under what may be regarded as a better organisation in another; and if the ungracious task were undertaken, errors in statement would inevitably be made. They would arise from the necessarily imperfect knowledge of the inner life of one as compared with that of another; but they might seem to proceed from the wish to support a foregone conclusion, or to show a want of generous appreciation of much work well done, and of the desire to have done it better.

But without crossing the threshold of any institutions it is possible from their published reports to illustrate and inforce what has already been said. Economy is the touchstone by which the management of a hospital may fairly be tested, and tried by this it will be found that wherever there is a strong secular administration the cost is low, and that it is high in almost exact proportion to the degree in which a sisterhood, or an association similar to a sisterhood, bears sway. If two hospitals at distant parts of the metropolis, and resembling each other only in this, that they both are under lay control are compared in this respect with the Children's Hospital, we arrive at the following startling results :—

	At the Westminster.	At the London.	At the Children's.
	£ s. d.	£ s. d.	£ s. d.
Cost of patient per day	3 7	3 4	4 9
Provisions in year for patients and attendants, per head . . .	17 13 11½	16 1 7	19 9 7½

These figures represent a weekly cost per head for provisions, not including wine, but reckoning beer of 6s. 9½d. at the Westminster, of 6s. 0d. at the London, and of 7s. 8½d. at the Children's Hospital.

The secretary of the Evelina Hospital was good enough to send the following reply on the 1st of January, 1877, to a question addressed to him on this subject :—

"The average cost for provisions (including beer) per head per week at this hospital is four shillings and tenpence (4s. 10d.)."

As a further element for comparison, it may be added that the weekly cost for provisions at the Children's Hospital, in which 468 out of 869 patients admitted in 1876 were under six years of age, is higher than the weekly cost per head during school-time at a first-class boarding-school for boys of the upper class of society, where the proportion of adults, as masters, servants, &c., was the same as at the hospital ; the ages of the boys ranging from eight to fifteen ; the boys all having meat twice a day, and most taking beer once, and most members of the household twice, in the day. This statement is made from actual inspection of the school accounts.

The details are also very singular, and not easily explicable ; for it is difficult to understand how the cost for meat should be £6 4s. 2d. per head at the Westminster Hospital, £7 3s. 11d. per head at the London Hospital, where 5,305 out of the 6,303 patients are adults, and where a large number of accidents are admitted requiring exceptionally high diet, while at the Children's Hospital, where no accidents are received the cost for meat alone is £9 3s. 10d. per head.[1]

It may be interesting to compare the three hospitals a little in detail with reference to the cost of provisions

[1] The reports of the Paris hospitals, which are all under the same system of management, furnish a clue as to what should be the cost of a hospital for children, as compared with a general hospital, and it will be seen that while the cost of both has greatly risen,

—a matter specially within the province of female management.

	Westminster H.	London H.	Children's H.
Beds permanently occupied	135.5	630	80
Patients admitted in year .	1,684	6,303	869
Adults engaged in service .	38	158	38

	£ s. d.	£ s. d.	£ s. d.
Cost of Meat	1,080 12 8	5,668 17 1	1,084 17 11
Bread and Flour .	250 7 5	1,306 4 1	163 3 10½
Milk	521 13 11	2,021 8 10	390 11 10
Beer . . .	76 14 11	1,037 3 2	94 17 5
Butter & cheese & eggs	232 17 2	1,497 6 7	254 11 8½
Tea, groceries, corn-chandlery . .	144 14 11	380 13 11	148 17 4
Fish	42 5 2	286 4 9	44 5 5½
Potatoes . . .	56 2 0 ⎫		
Green vegetables, and extras . .	136 3 3½ ⎭	472 12 6	117 10 11½
Additional, not included above, for officers and servants	497 13 6 ⎫ 57 17 4 ⎭		
	3,097 2 3	12,670 10 11	2,298 16 6
Washing	383 10 0	827 15 4	414 11 0
Coal, coke, and gas . .	549 11 0	2,529 2 4	646 7 8

the proportion still holds good almost as markedly as twenty years before.

Average daily cost per patient in General Hospital.		Ditto for food and treatment.	
Fr.	Cent.	Fr.	Cent.
In 1849 1	99.78	1	12.99
In 1869 2	53.62	1	60.58

Average daily cost per patient in Children's Hospital.		Ditto for food and treatment.	
Fr.	Cent.	Fr.	Cent.
In 1849 1	28.68		93.12
In 1869 2	6.29	1	22.46

Cent.

The children's hospitals were less costly in 1849 by 71.10
in 1869 by 47.33
The food and treatment were less costly in 1849 by 19.87
in 1869 by 38.12

These figures are drawn from the official *Compte Moral de l'Administration de l'Assistance Publique,* published annually by the Conseil Général de l'Administration des Hopitaux, &c. de Paris. All hospitals are of necessity more costly than they were twenty-five or thirty years ago, owing partly no doubt to general increase of prices,

The following memoranda will help to prevent misunderstanding.

The cost of board at the Westminster is calculated on the footing of board for 40 adults, since meat lunch is provided daily for 6, and bread and butter and cheese lunch for any member of the medical staff. Coal and gas are also provided for the medical school. At the London—tea, sugar, and butter are not provided for the patients ; nor washing, except of their uniforms, for the 113 nurses. Gas is used in cooking, and the washing is done in the laundry. The cost per head for coals, gas, &c., is however £5 9s. 6d. per head at the Children's Hospital ; £3 2s. 11d. at the London, notwithstanding there being two steam-engines at work there and only one at the Children's Hospital. It would be unfair to institute a comparison between the cost of washing at the London Hospital, where there is a laundry, and at the Children's Hospital, where there is not. At St. Bartholomew's Hospital, however, where there is also no laundry, but where there were 519 occupied beds in 1875, the cost of washing was £980 5s. 3d.[1] as against £414 11s. 10d. at the Children's Hospital, where the occupied beds were 80.

These facts comment on themselves. What need is there for further remark ?

The principles hitherto laid down apply to the government of all hospitals, but the details on which

---

but there are no causes specially affecting hospitals for children which should make the daily cost per patient at one general hospital be ·40 ; at another ·43 ; and at a children's hospital ·57 ; that in other words, it should be the costliest in the proportion in the one case of 13, in the other of 17 per cent. ; or that it should be possible to maintain three beds at the London Hospital for little more than the cost of two at the Hospital for Sick Children.

The cost at the Westminster Hospital is that obtained by dividing the sum of £65 8s. 8d., the estimated cost of a bed, as given by the courteous secretary, by 365. The estimate for the London rests on the authority of the house governor, after deduction of the cost of out-patients, and includes the whole of the expenditure, even for such matters as printing and stationery, and for ordinary repairs, needed for the maintenance of the building, for furniture, &c., and everything except that which is really extraordinary. A similar deduction for out-patients, according to the secretary's estimate, and for the expenditure which figures as extraordinary in the report of the Children's Hospital, yield the figures above given, or a cost of £86 per head annually. The published statement of the cost at £74 10s. 0d. per bed yields a daily expense of 4s. 1d., but similar deductions applied to the London Hospital would leave the comparative cost of the two unchanged.

[1] As appears from the account of income and expenditure rendered annually to the Charity Commissioners.

we are now about to enter have reference especially to hospitals for children, and even these details must of necessity be to a great degree fragmentary, and are suggestions offered for consideration rather than regulations laid down as of universal application.

PART II.

The first question that comes before us has reference to the AGE OF THE PATIENTS, and the determination of the limits below and above which their admission is inexpedient.

The minimum age for admission into a children's hospital should be fixed generally at two years. Below that age children are to all intents and purposes infants. They often have not cut more than half their teeth, in many instances they need to be fed from the bottle; they are too young to be cleanly, and need the same attention as a babe of a few months old. They often can be soothed only by nursing in the arms, and require so much individual care that two children of eighteen months old in a ward would require one additional day and one additional night-nurse to tend them properly. Here of necessity comes in the question of expense.

If this attention is not given, the infant will disturb the ward by its cries by night as well as by day; and moreover the neglect of individual attention to young children is sure to be followed by a great increase in their mortality; the children dying not so much of the disease as from the want of individual care.

The majority of the diseases of early childhood are either very acute, as inflammation of the lungs, tending, if not cured, to a rapidly fatal end, and requiring the personal attention of almost one nurse to each child; or else they are diseases dependent on malnutrition and its results, and are of a scrofulous or tuberculous character, as water on the brain, and the various forms of consumptive affections, and surely tending to death. The good accomplished by their reception would be small, the cost immense; the scandal of the high death-rate, of which the public would not understand the reason, would be immense too.

It is therefore most undesirable to make the admission of children under two years of age otherwise than exceptional; but the medical officers should have a discretionary power to admit such children for important medical or surgical reasons. To prevent any misuse of this power, however, the name of the patient, with the reason for his admission, should be entered in a book to be laid before the house or managing committee at each meeting.

The reasons for restricting the age for the admission of children to those who have not exceeded that of ten instead of, as in some children's hospitals, twelve years, though neither so cogent nor so obvious as those which limit their age in the other direction, are yet very weighty.

After the age of ten years diseases lose to a very great degree the peculiar characteristics which they present in earlier childhood, and approach more and more to those of the adult.

Children after the age of ten are readily received into the general hospitals. Any provision therefore for their ordinary reception into children's hospitals is unnecessary; and by their frequent admission the beds which should be filled with the *special* diseases of childhood are diverted from their real purpose.

E

In a good many instances, both boys and girls, especially the latter, at and about the age of twelve years, and even earlier, have begun to undergo the changes of approaching puberty. If it became at all customary to admit boys and girls into a children's hospital after the age of sexual consciousness has come—and that comes very early in London children —there will be very great difficulty indeed in maintaining that purity of the moral atmosphere around the children which is as important in one respect as the ventilation of the wards is in another.

The same power of exceptional admission exercised by the medical officers in the case of children under two should also be, with the same restrictions, vested in them in the case of those above ten.

From the Patients, the transition is a natural one to THE NURSES.

It has sometimes been imagined that in a children's hospital the nurses should as a rule be younger than would be either necessary or desirable in a hospital for adults. The opinion is very plausible, and in the case of those who have charge of convalescents is perhaps not without foundation. But a child when really ill cares less for the brightness of youth in its attendants than for the gentle patience which comes at a somewhat later age. The great reason, however, for not engaging nurses, or even admitting probationers, at eighteen or nineteen, is not only that less dependence can be placed on them than on those who are older, but that young girls do not bear the fatigue and confinement of nursing, and especially of night nursing, well; that till the age of perfect, well-developed, full-established womanhood has been attained, the strain upon the constitution is often more than they can bear. Twenty-one is probably as early an age as it is wise for any woman to enter on the duties of a nurse. It would also save much disappointment both to the nurse and to the superintendent of nurses if all young women underwent a medical examination before entering on their new duties. At least as much care should be taken in the admission of a recruit into the army of charity as into the army of any sovereign.

The standard of education for the nurse ought not to be fixed too high. She should be able to read writing as well as printing easily; she should be able also to write legibly, since she may have to keep memoranda with reference to her patients for the information of the head-nurse, or of the resident medical officer; and she should be able to do easy sums in the first four rules of arithmetic; or, in other words, should have a moderate elementary education.

For head-nurses the standard should be fixed much higher; how high need not be defined here; but she certainly ought to have the whole technical knowledge of an experienced nurse before she should be allowed to superintend others; should be able to understand prescriptions, and should have served for at least six months in the dispensary.

The time of TRAINING for a probationer before she should be considered competent to act as a nurse in any hospital ought to be not less than one year: and no person should be appointed to the office of sister or head-nurse without having previously filled the post of nurse for one year after the completion of her term of probation.

It would be extremely desirable that in the case of a head-nurse appointed to hold office in a *special* hospital, such as a children's hospital, she should have passed at least six months, either of her training, or of the term during which she was discharging the duties of a nurse, in some *general* hospital.

But it may be asked, What is meant by training? It is something which, as a rule, neither hospital nor private nurses have had in any true sense, but yet it is regarded as so important that the question of how to supply it has engaged the attention of the General Medical Council.[1]

In the majority of hospitals training has had no other meaning than this: that a woman desirous of engaging in nursing is admitted at low wages, or at none at all, or sometimes paying for her board; that she is set by the head-nurse to help, first in small things, then in those that are more important; and that thus she by degrees picks up a measure of practical acquaintance with how to manage the sick, how to arrange their bed, how to bandage a limb, and so on. If she is very acute and observant, and painstaking, she may at last arrive at a degree of dexterity in the technical part of her duties which many a surgeon might envy, and may acquire an insight into the probable nature and issue of a disease which would do no discredit to the experienced physician. This standard has actually been reached by some of the old Sisters in our large hospitals.

But these cases are exceptional; and even the ablest self-taught people have deficiencies in knowledge, and faults in character, which proper education would prevent. Hospital nurses as a rule are not trained, are not taught. The Nightingale nurses who receive instruction at St. Thomas's, and the Liverpool

[1] See *Minutes of Medical Council*, vol. x. pp. 169, 189.

nurses, of whom those at the Westminster Training School are an offshoot, are even now almost the only exceptions.

To train a nurse really, implies the co-operation of some at least of the medical staff of the institution, and the active share in the work of the superintendent of nurses, who ought herself to be the best nurse in the hospital.

Some small amount of elementary knowledge of the structure of the body, and of the signs and tendencies of disease, should be given orally by the doctors, and the mode of teaching should be as far as possible friendly, conversational, catechetical, and as little as possible purely didactic. The nurse should be taught what to observe, and why in one disease one set of symptoms is to be specially noted, and why in another a different set.[1] She should be instructed in the simple tests of the urine, and in the use of the thermometer, as well as shown how to make a poultice, and how to apply leeches; but the wisdom of her instructor will appear most in the endeavour to teach but little, and to teach that little well. She should learn to bandage and to dress wounds under the house surgeon; and should not be left to learn everything by practice, but should be shown why one way of doing a thing is right and the other wrong.

The superintendent of nurses will have a far larger and more active share in the good training of the nurse, than any or all of the medical staff. She should really be what in a religious order would be termed the mistress of the novices; and, if she realizes the im-

---

[1] To do this was the purpose of a little book, *How to Nurse Sick Children*, published years ago, and at first anonymously. Its object was a very humble one, but even in this it does not seem as yet to have been superseded.

portance of the work she has to do, will have neither
the leisure nor the wish to indulge the foolish ambi-
tion of governing everything ; a wish which, if grati-
fied, is sure to end in the mismanagement of everything. There is no detail of the nurse's duty which
she should not every now and then supervise herself,
instead of leaving all to the oversight of the head-
nurse, how competent soever she may be. As the
doctors, too, must teach orally sometimes, so must
the superintendent of nurses do the same in what
there can be no difficulty in practically marking out
as the specially womanly part of the work. When
this is done, but not till then, the nurses in a hospital
can be said to be properly trained.

But while thorough training is essential to make
a thorough nurse, the question must not be passed over
unnoticed whether any smaller amount of education
will suffice for those who have no intention of following the occupation of nursing permanently. The idea
that such a plan was feasible was generally entertained
when the Children's Hospital was founded; and the
" Training of Nurses for Children " still appears at
the head of its reports as one of the primary objects
of its establishment. The idea, however, has not
been realized at all ; apparently because, in the first
place, young women seeking for occupation in the
nursery did not see the importance of a small amount
of knowledge about the diseases of children, and had
not the money to spare for a three months' residence
in the hospital ; and, in the second place, because
ladies did not themselves appreciate its value, and did
not in any way co-operate by sending young women
from their own nurseries to receive instruction.

It is very probable that these difficulties may be

insurmountable, since the causes which rendered the project abortive from the first are likely to continue always in operation. Even though this should be so, there will still be some—and those chiefly of a higher social class than that from which paid nurses are usually supplied—who will wish for various reasons to acquire some knowledge of disease, and especially of children's diseases and their management. Such will be the wives and daughters of clergymen, or others whose duties bring them much in contact with the poor; such those who are intending to engage in missionary work, or who are about to settle in India, or to live in the remoter parts of the colonies; and such too will sometimes be young ladies who, having run through the gaieties of one or two London seasons, find out all at once, like the little girl in one of Punch's sketches, that "the world is hollow, and my doll is stuffed with bran," and so take up for a time with a new form of dissipation. Of this last class, the fewer who come to learn nursing the better, at any rate for the institution which receives them; though it must be admitted that now and then the realities of life steady them, frivolity gives place to earnestness, and they throw themselves heart and soul into their new work.

Three months spent in a hospital, under the watchful eye and with the helpful care of the superintendent of nurses will teach an educated woman very much indeed of what it is most important for her to learn. But if she is not to disarrange the well-working of the hospital by the introduction of the mischievous aristocratic element, she must take the same place as the others, and the distinction between lady pupils and probationers must be unknown.

"Lady pupils," says Miss Lees,[1] herself no small authority in such matters, "should pay a fixed sum for board and instruction, but in all other respects be treated the same as the other probationers."

There will be no hardship in this, if that which has been so strongly contended for some few pages back— the absolute exclusion of caste—be the principle on which the institution is conducted. In that case she will just take her place by the side of others who are doing the same work with herself, the only difference being that she comes as a learner, whilst most of the others already know their business.

It is at the present day almost impossible to avoid THE RELIGIOUS DIFFICULTY altogether. In a children's hospital the duty of the head-nurse, probably also occasionally of one of the other nurses, will be to read some little prayer, morning and evening. It would be very undesirable that any one whose religious opinions raised a question in her mind as to the propriety of doing this should be appointed a nurse ; and therefore, while no religious restriction should be allowed to interfere with the admission of any one temporarily as a pupil to train, it would probably be a sound rule to require that all those who are permanently employed about the children should be Protestants.

And here, as the question of religion has been introduced, is perhaps as good a place as any to inquire how far religion should be made a predominant element in the daily life of the hospital nurse. There is no work, be it what it may, which religion does not elevate, does not beautify, does not sanctify. But to postpone the duties of the nurse, to throw the

[1] *Handbook for Hospital Sisters.* 8vo. London, 1874, p. 33, note.

machinery of daily routine out of gear in order to secure the attendance of the nurse at morning and at evening service in the hospital chapel, is a mistake, and a grievous one. And here we fall back again on the wise words of St. Vincent de Paul. He tells the sisters [1] that to wait upon the poor everything else should be left ; even prayer and the mass ; for as he constantly repeated, it is "leaving God for God." "Do you fancy," he said, "that God is less reasonable than an earthy master would be, who, having told his servant to do one thing, before he had had time to obey his command tells him instead to do something else ? The master would not find fault with his servant for not doing in this as he was bid at first; he would indeed be only the better pleased with him for it."

Nor indeed will it be out of place to borrow an illustration from the history, or, as some would call it, the legend, of St. Elizabeth of Hungary, whose time was divided between prayer and works of charity. It was vouchsafed to her to see the Lord she loved so much and served so well, not in the hour of meditation or of prayer, but when she had brought home the leper child, and laid him on her bed, and left him to meet her husband angered at her lowly work, she found on thence returning the leper child no longer there, but the Babe from Bethlehem.

And so if we can but teach ourselves and others to realize the poet's words, and really feel that

> " The trivial round, the common task,
> Would furnish all we ought to ask ;
> Room to deny ourselves, a road
> To bring us daily nearer God,"

the work will be better done ; we shall be treading

---

[1] Maynard, *op. cit.*, vol. iii. p. 251.

closer in our Master's footsteps; be more likely to meet Him in it, than if we postpone work to worship, or exhaust in long meditation or in prayer the strength which is needed for our active duties.

The RELATION OF THE NURSES TO THE HEAD NURSE OR SISTER should be that of a private soldier to his non-commissioned officer; not that of a domestic servant to her mistress; and the class distinction which one hears in some institutions between the *ladies* and the *nurses* should find no place in a well-ordered hospital. It will probably be the case that the promotion of a nurse to the position of head-nurse or sister will not be of frequent occurrence in the same hospital, and that in the greater number of instances the post of sister will be best filled by some one who has undergone her training, and has occupied the subordinate place in another hospital. But at the same time it is very inexpedient to take away from the nurse the possibility of promotion, if not in her own hospital, at any rate in another, and to destroy that stimulus to improvement which the power of rising by merit yields to every one in all the occupations of life. The existence of a governing caste from which others are hopelessly excluded is fraught with danger to the good working of any institution, and constitutes one, and certainly not the smallest, of the drawbacks inseparable from a sisterhood. The mere fact of the one class being paid and the other not receiving money, affords no justification for such an arrangement: for, as Miss Stephen truly says in her valuable book,[1] in which the comparative merits of sisterhoods and of lay agency in works of charity are very fully and ably treated,—"The fiction of gratuitousness is

[1] *The Service of the Poor.* Post 8vo. London, 1871, p. 239.

founded on nothing more real than the difference between money and money's worth, and it is a fiction which almost inevitably tends to make nominally unpaid workers deceive themselves, and take credit for a disinterestedness which they do not possess. I cannot help thinking that there is an unconscious dishonesty about this arrangement which is characteristic of a system whose foundation is aspiration rather than fact and typical of that indiscriminate claim to the credit of self-sacrifice put forward on behalf of sisters of charity as a body in more important things, without inquiry, or in spite of evidence, as to the balance of loss and gain involved in their position. It is necessary to dwell a little upon this divergence between the name and the reality as regards payment, because the plea of disinterestedness is so constantly put forward as a sufficient guarantee for the excellence of the motives and character of all the members of these nominally unpaid bodies; and discredit is thrown by comparison upon secular nurses and others, working simply, and without pretending to any merit, for honest wages; such women are called mercenary and hirelings, and we are asked how we can expect devoted service from those whose object is to gain money. I say, in the first place, that if earning one's living were incompatible with self-devotion, there would be but little self-devotion in the great mass of the charitable orders, for I believe that ninety-nine out of a hundred of their members to be honestly earning their living; and, secondly, that it is a fallacy whose injuriousness is partially redeemed by its transparency to confound the conditions of any service with the motives of those who belong to it."

The extract is a long one; but he were a bold man who would venture in the present day to write these truths about women and their work; and hence the alliance of a woman who dares to *say* what many men *think* is too valuable to be neglected.

It is especially for the sake of doing away with class distinctions that, independently of its great importance on the score of economy, it is desirable that there should be a common table for nurses and sisters. By nothing is the distinction between the two made more marked than by their meals being provided at different hours, in different rooms; and this, even though the diet of the two classes were the same, which it commonly is not. The mixing, or the not mixing with others in ordinary intercourse, is one of the great marks of persons belonging or not belonging to the same grade of society. It is on this principle that we do not ask our shoemaker or our baker to dinner, and that those of us who belong to the middle class do not expect to be invited to the table of princes or of the sovereign. The story goes, that when Louis XIV. wished most effectually to rebuke the insolence of the nobles of his court to Molière, he made the actor sit down to dinner with him. The tale indeed is mythical, but it serves as an illustration just as well as if it were true.

It does seem somewhat strange, that they who profess to have left all for the service of God's poor should fancy it essential to the well-working of a hospital to maintain rigidly the social distinctions of the outside world; and one can scarcely suppress a smile at finding oneself taken back to the time of the Pharaohs, when "they set on for him by himself; and for them by themselves, and for the Egyptians by themselves."

"It is one thing," says Miss Stephen,[1] "to wash the feet of the poor in good company, and quite another, and to many ladies a much more trying thing, to take an undistinguished place side by side with one's social inferior : for instance, to travel third class, not in the dress of a Sister of Charity, but in that of an ordinary third-class passenger ; or to sit down to meals with paid nurses in an ordinary hospital." In this, however, there is nothing but what is done in every Roman Catholic convent, although the community is composed of women of very different classes. There is a common refectory, and meals are taken at the same time, as far as the work to be done admits of it; while the food is absolutely the same for all. At the top table sit the Superioress, the office-bearers, and the professed sisters. Below them are seated the novices and the lay sisters. Their rank before they entered the community has no place within the convent walls, nor should it have in a hospital. By the opposite plan the opportunity for that cheerful, friendly intercourse by which so much good is often indirectly done, is thrown away to gratify a sort of aristocratic feeling; and real friendliness between the nurses and the sisters can scarcely spring up if the two classes associate only in the discharge of their duty.

But the principle of a common refectory has great value also as far as economy is concerned. Let the arrangement for meals be made as perfect as possible, yet each must be partaken of in two divisions : one-half of the nurses remaining in the wards, while the other are at breakfast or at dinner. If to this it is added that separate hours must be kept for the

_____
[1] *Lib. cit.* p. 236.

night nurses as distinguished from those on day duty, it will be found that altogether some eight distinct meals will have to be provided for the hospital staff. The lady superintendent and the house-physician and house-surgeon must dine separately, if for no other reason, yet, because till toward evening they cannot be certain of being at liberty ; and the arrangement for a common table for the resident officers, while it undoubtedly diminishes the amount of labour required, has yet its drawbacks, and cannot be universally adopted.

If to this unavoidable demand upon the resources of the hospital and the labour of the servants separate meals are introduced for the head-nurses; often at different hours, often composed of different dishes, and always served in a different room, it becomes necessary to tell off a distinct attendant to wait upon the sisters. Moreover, fires are kept burning, or gas is consumed for cooking, twice over ; the work of the domestic staff becomes never ending, and the funds of the hospital are squandered for the sake of pandering to the class prejudices of those who must lay them quite aside if they leave their original station in society in order to be of any real use in hospital work. Simplicity, as already said, is the motto which should be written everywhere, stamped on every arrangement in a hospital. It may be feared that all of us have sinned more or less in forgetting it. Polished floors and glazed bricks, or Parian cement, may be justifiable for hygienic reasons ; but it may be doubted whether that spirit of luxury which has invaded our own homes does not too much inspire us in our charitable work ; whether our own drawing-rooms are not too much in our minds, and too little the homes of the poor, to which it is the object of a hospital to return them healed in body, unspoilt in mind, as quickly and as cheaply as possible.

A good, simple, uniform dietary, and a common refectory for all the head-nurses and nurses, should be a rule without exception in every hospital. The refectory should not be the day-room, but there should be a separate room for the rest, and recreation of those who are engaged in nursing the sick. The arrangements of the refectory should be under the care of the housekeeper, those of the day-room under the rule of the lady-superintendent ; but a day-room maid, as she is termed, whose duty it is to answer the sisters' bell, and to wait upon their bidding, is a solecism on the staff of a hospital which cannot be too soon suppressed, even if such an attendant did not cost the institution a single penny.

It is almost impossible to lay down any fixed rule as to the number of nurses required in any hospital. In general hospitals this is governed to a great degree by the nature of the cases admitted : and especially by the proportion of accidents and severe injuries commonly brought to it. Thus we find that at the Westminster Hospital there is one nurse to every six beds ; at the London, which, owing to its nearness to the docks, receives more serious accidents and injuries than any other hospital, the proportion is 1 to 5·5 ; and this is inclusive of probationers as well as of trained nurses. The proportion of children under 12 is about 1 in 8 in the Westminster, 1 in 6 in the London Hospital. It is probable that in proportion to the number of patients children's hospitals require more nurses than hospitals for adults ; and the practice in the Paris hospitals, in which this is not, or at any rate very recently was not, the case, cannot be appealed to, since most who are acquainted with the subject will agree in thinking that the French hospitals generally are under-nursed.[1] But when we find that at the Children's Hospital in Ormond Street, into which no accidents are admitted, 19 nurses and 5 sisters have charge of 80 patients, or in the proportion of 1 nurse to 3·3 beds, the impression is not unnatural that the nursing staff is disproportionately large.

A difficulty indeed which is felt in every hospital, and has a tendency to swell the number of nurses beyond that at all times required, arises from the necessity of providing for occasional emergencies. It has been proposed to secure a reserve for such occasions, by employing the extra nurses when not needed in private nursing, by which means the expense of their maintenance would be borne by the families who employ them, and some money even would accrue to the funds of the hospital.

This proposal shows such a strange want of perception both of the nature of the difficulty and of the way to meet it that it might be left unnoticed but for a certain air of plausibility which deceives people who do not give themselves the trouble of thinking twice.

---

[1] The only authoritative statement on this subject, as far as the writer knows, is to be found in the great work of Husson, *Étude sur les Hôpitaux*. 4to. Paris, 1862. He gives at p. 175 a table showing the number of attendants per bed in the different hospitals at different periods. For the year 1862 the proportion was—

Hôtel Dieu	1 to 8·49
Pitié	8·45
Charité	9·0
Lariboisière	9·62
Enfants Malades	9·22
Sainte Eugénie	7·87

But these figures overstate the number of those actually engaged in attendance on the sick, since they include also the *serviteurs des salles*, whose duties are to a great degree those of servants, not of attendants on the patients.

An emergency is something sudden, unexpected, impossible to be foreseen; the sudden outbreak of diphtheria, the necessity for opening the wind-pipe; the need of the presence of a nurse day and night for one or two days to watch a patient after an operation. What advantage, when such a case arises, will be the knowledge that a certain number of nurses are scattered here and there in different houses, whence they can be recalled only after due notice given? nor always even then, for if the patient whom she is attending is very ill, and both he and his friends have become much attached to and very dependent upon her services, it will be practically impossible to summon her away, because she is wanted at the hospital. The reserve will still be needed on the spot, ready to give immediate help, or that reserve is valueless; till things change their nature, until the superintendent of nurses becomes gifted with a prescience not bestowed on ordinary mortals.

But, supposing for an instant that all nurses could be recalled if wanted on twenty-four hours' notice, it would not be at all desirable for the good working of the institution that they should be employed outside the hospital. One of the complaints already mentioned as having been alleged against Roman Catholic religious communities in connection with hospital work was: "The occasional arbitrary change by the superioress of the community of one sister who had been in a hospital for another less experienced, or to whom, at any rate, the ways and arrangements of the institution were strange;" and similar complaints have been made of changes in the nursing staff of at least one hospital nursed by a Protestant sisterhood by the medical officers of the institution. The nurse sent

F

out would, for the sake of the reputation of the hospital, always be one of the best and most experienced. Her selection must of necessity rest with the superintendent of nurses; for when private families send for a nurse they want her at once, and cannot wait for consultation between the superintendent and the principal medical officers, while her authority would of necessity override that of the house-surgeon; and its exercise on these occasions would be a fruitful source of quarrels between herself and him. The displeasure of the physician or surgeon would not be unreasonable if, on making his rounds, he found that the sister on whose judgment he relied, or the nurse who had dressed a wound with singular gentleness and skill, had disappeared for an uncertain time when he thought she was most needed; and he would scarcely be mollified by hearing afterwards that the family, one of whose members she had nursed, were so delighted with her, that in testimony of their gratitude they had become life governors of the hospital.

But the influence on the nurse herself of such change from hospital work to private nursing would be in the highest degree injurious. Her interest and her affections should centre in the hospital, with its regularity and simplicity, so long as her duties lie there. It is not indeed without reason that those among the Roman Catholic nursing orders, who are engaged among the poor, are precluded by their rule from attending on the rich. The change from the ward, from making the patients' beds, from the plain food, and simple ways, to the grand house of the wealthy, to being waited on by others, to costly food and luxurious living, would spoil all but the best and wisest of their class; and we have no right to expect from any one in any station more than a certain rather low average of wisdom or of goodness. The saving of the nurse's keep and wages for a few weeks, and the advertisement of the hospital which her good qualities may have furnished, are bought too dear at the risk of impairing that simple earnestness in her ordinary work which it costs so much pains to cultivate and so much watchfulness to retain.

The only way in which the desired end could be attained would be by the establishment in connection with a hospital of a training school and home for nurses, after the pattern of the Liverpool Institution, or of that in connection with the Westminster Hospital. There the hospital pays £8 per bed annually for nursing, while the home is debited with a charge of £25 a year for each nurse's board. Such a home could be drawn upon in an emergency for the supply of extra nurses, since there would always be some there who had just returned from one engagement and were waiting for another; and there would be no difficulty in settling the details of the arrangement between the two institutions for the payment of extra services. The nurses ordinarily employed in the hospital would not be changed, but would remain a fixed number.

The head of the training institution should have no power to substitute a new nurse for one already working at the hospital without the consent of the medical officers and of the superintendent of nurses, and the authority of the director ; while she should be bound to change any one concerning whom those officers made complaint, either for the manner in which she performed her general or her medical duties ; with reference to which latter, the representations of the medical officers should always command attention.

It is needless to go into further detail here as to the working of the two institutions so as to ensure the harmony of their action.   It is evident, however, that the difference is fundamental between this plan and the proposal which has been the subject of comment. The Home and Training School, whether intended for lay-nurses or for a religious community, must have a distinct organisation, its own separate rule and its own head, over whom the superintendent of nurses would have no control whatever ; and this even though it were never so closely connected with the hospital, and even supported out of the hospital funds.

Before the public can be appealed to successfully to support an undertaking which, though good in every way, would yet require very careful organisation and skilful management, they would need to be satisfied that the present work of any institution was well and wisely done.   There is no more hopeless form of incapacity than that which shows itself in the readiness to embark in some new enterprise while old work is undone, or ill done, from want of knowledge or from want of will.

Pending the establishment of a home and training school for nurses in connection with a hospital, there are yet some simple means by which, especially in a children's hospital, the nursing power can be economized, and provision made to meet, partly at least, some emergencies.

1st. Since, at children's hospitals, it is always customary for the out-patients to be seen in the morning, the supernumerary nurses might, both with advantage and economy, be employed in the out-patient department. It would rarely happen that all supernumeraries would be called upon at once, and, even if that did occur on some occasion, it would be much easier to provide attendants from without for the out-patients than nurses for the sick within it.

It need scarcely be said that it is not suggested that the supernumerary nurses should form a class apart, but merely that there should be a sufficient number of nurses to afford relief to those who are overworked, or to meet any accidental demand for extra help. The out-patient work would occupy about four hours at the most, and a nurse unfit for the time for work in the wards might well perform those duties, and, if necessary, have the whole remainder of the day for rest and relaxation.

2nd. All children whose cases are not severe should be removed into convalescent wards. Those who are well enough to leave their bed would spend the day in the convalescent day-room ; where as in an infant-school one nurse would readily take the oversight of fifteen or twenty children, while many suffering from chronic ailments, and who were unable to leave their bed, would yet require a far less amount of nursing than would be called for by the acute and more serious cases.

3rd. The under-nurses should not be appointed so permanently to one ward that they could not be moved about to take duty where they were most wanted. What should we think of an officer, who, in face of the enemy, kept the same number of men at each outpost ? It does not need that one should be a military man to say that he would concentrate his men where they were most wanted, and even with-

draw them altogether from one point which he felt
secure, to send them to another where danger was im-
minent. And yet the nurses are supposed to be so
strictly attached to one ward, that while the attendants
on the sick in Charity Ward have nothing to do, those
in Faith and Hope are overwhelmed with fatigue. The
limitation of a nurse's work is sometimes carried so
far that she who has charge of one side of a ward
may be sitting idle, while her fellow-worker who has
the opposite side to look after may be over-burdened
with bad cases.

The exact rota of a nurse's duties, the hours for
work, the intervals for rest and recreation, for air and
exercise, must vary so much that it were idle to
attempt to lay down any general rule on the subject.
But yet, to whatever extent it is possible to have the
hour for the medical visit fixed for the morning, so as
to leave the afternoon comparatively free, that object
should not be lost sight of. In a general hospital
with which a medical school is connected, an almost
insurmountable objection to such an arrangement
arises from the impossibility of the student who has
lectures to attend, and dissecting to practise, following the visits of the physician or surgeon before one
or two o'clock in the afternoon.

Attendance in a children's hospital, however, comes
at a late period of a student's career, usually indeed
after he has passed his examinations, and at a time
when he is seeking to supplement his general knowledge by special study. To him, therefore, the morning hour would offer no obstacle, while all who are
conversant with the diseases of early life must acquiesce in the extreme importance of the medical
visit to children whose ailments are either acute or

severe being made early in the day. In surgical cases, too, nothing can be more distressing nor more injurious to the child than its disturbance twice in the day ; once for the necessary morning visit of the house surgeon, and again later in the day when the surgeon goes his rounds, and the bandages have once more to be undone, and the wound again exposed for his examination.

As far as the general conduct of the institution is concerned, it is of great moment that the real work in a children's hospital should be done in the morning, the new diet ordered, the new medicine prescribed, the new dressing to the wound applied ; and that the nurses should have an interval of rest after dinner, till their evening duties call once more for their attention. Valuable, inestimable even, as the services of the medical men may be, a large institution like a hospital must be worked for the general good, not for individual convenience, so long of course as the latter is not ignored, and as regulations from which some dissent are not enforced without sufficient reason.

There are two different plans according to which the arrangements for day and night nursing in hospitals are regulated, and each has its own advantages. According to the first, the night nurses form a separate class, who never perform day work, but coming on duty from eight to nine in the evening, go off duty at seven or eight in the morning, and handing over the patients to the day nurses, and giving their report to the sister in charge, spend the day partly in rest, partly in recreation, and resume their work when night comes on again. This seems to be the arrangement, which, as far as the health of the nurses is concerned, works best, and women, especially after the age of thirty or thereabouts, appear to be capable of undergoing it for years. At one time these night nurses were an inferior class, receiving lower wages than those who nursed by day; but this mistaken practice has long since been given up in all hospitals, and there is now no longer difference of pay between the night and day nurses. The drawbacks from this plan are that the interest of night nursing is much smaller than that of nursing by day, though the responsibility is really greater and the qualifications called for are of a higher kind, while the elements of individual attachment to the patient, and all that womanly tenderness which plays so large a part in the good nursing of the child, is scarcely evoked. Moreover the night nurse seldom or never comes into personal contact with the doctor; and all those words of encouragement, of praise for duties well performed, which each of us in our several stations values so highly, it rarely falls to her lot to hear. At St. Thomas's Hospital some special arrangements are made to do away with this evil, and to give the night nurse on certain days the opportunity of meeting the doctors.

The drawbacks above referred to have led to the general adoption of a system of alternate day and night nursing among the attendants. The mode of rotation varies. Many years since the plan at St. Bartholomew's Hospital was as follows: Each ward of twenty-four beds having three nurses besides the sister, all were on day duty, and each every third night was on night duty too, thus having on these occasions thirty-six hours of continuous duty. This most unreasonable demand upon a woman's strength has long been a thing of the past, and is referred to here merely as the first rough attempt to solve the problem of day and night nursing. The present arrangement in most hospitals is that each of three nurses in a ward commonly has a six weeks' spell of day nursing, and a three weeks' spell of night duty, but young women are often much tried by this; and many find that it takes two or three days before they can turn day into night, and succeed in getting refreshing sleep after a night's watching. The subject is a most important one, and calls for more careful examination from a medical as well as an administrative point of view than it has yet received.

As a means of saving the nurse from over fatigue, or from the performance of duties which may be supposed to be beneath her, a practice has been adopted in some hospitals of attaching to them a subordinate class of assistants under the name of *Ward-Helpers.* This plan, however, is by no means to be commended. From an economical point of view the inferior class of persons are little less expensive than nurses, the only difference being in the rate of wages, and hospital waste and extravagance always take place in what may be termed the *unknown* quantities, not in the ascertainable figures of salaries and wages. The ward-helpers, moreover, being an unskilled class, cannot be available for nursing the sick, but are in reality merely the nurses' servants, having the same relation to them which a kitchen-maid has to a cook in a private family.

The amount of work to be done in the wards other than the care of the sick varies greatly in different hospitals, according to the peculiar nature and structure of each. No one would propose that the nurses

should scrub the floors, or black the grates, or do any
of that work which is commonly known as menial,
and the performance of which would render them less
apt in handling the sick with gentleness, or in performing all the operations of bandaging a limb, or of
dressing a wound with delicacy. All the rougher work,
however, admits of being strictly defined, the hours
for its performance can be fixed, and an arrangement
come to between the director, the lady superintendent and the housekeeper, as to the person by
whom such work shall be done, and as to the authority with whom it rests to enforce its performance.
The sweeping and dusting of the ward, the responsibility for its general cleanliness, neatness, and order,
must rest with the nurses themselves, and if the
blame for unwashed mugs and cups, for slovenly bath-
rooms and ward-kitchens can be shifted from themselves to some other person there will be an end to
all discipline. "I told the ward-helper to do this, or
that," will be the all too ready excuse for indolence
or carelessness.

The appointment of these servants to attend on
them fosters too the mistaken notion of the peculiar
sanctity of nursing, as though the ordinary occupations of woman were beneath that high vocation, and
the hospital entertained, at £25 or £30 a year wages,
not a set of honest, respectable, kindly women, to do
woman's work in the wards, and especially to nurse
the sick, but rather "a troop of shining ones," as
honest John Bunyan calls them, whose celestial
character would be degraded by the long broom and
the duster. There is a kind of Pharisaism to which
the half-educated are only too prone, and which inter-
feres grievously with the reality that should pervade

all hospital work. It would be far better to impress upon them two lines of the pious doggerel of George Herbert, which form part of the warp on which he here and there has woven gems of rarest beauty, like pearls upon a coat of frieze :—

> " Who sweeps a room but for Thy sake,
> Makes that and the action fine."

Before leaving the subject of nurses there are two matters concerning which a word may not be out of place. Nurses every now and then knock up temporarily from accidental overwork, such as the presence of an unusual number of severe cases at one time in the wards. They lose their appetite and strength, and even when the hygienic conditions of the building are good, they suffer from what is known as hospital sore-throat. If when in this state they go into the country for a short time they speedily recover, but the machinery of the institution is often more or less disarranged by their absence, and the hospital is put to expense to find temporary substitutes ; while they themselves may not always have friends in the country, nor be able to afford the journey thither even if they have. In those hospitals to which convalescent asylums are attached, there should be no difficulty whatever in sending any nurse, whose health seems failing, to the purer air and lighter duties of the convalescent hospital. If that plan of interchange of duties in the ordinary hospital work is carried out to which reference has already been made, and which common sense would seem to dictate, there ought to be no difficulty in its extension to the convalescent hospital. That which chiefly stands in its way is a strange kind of mutual jealousy between the heads of the two institutions. If this were laid aside, and if they co-operated in the one common object of doing the greatest amount of good possible, the convalescent hospital would be of as much use to the nurses as to the patients.

The other matter which will likewise tend to the real advantage of the nurses is the establishment of some sort of nurses' fund in connection with the hospital. This fund should be entirely distinct from a pension or superannuation fund. Faithful service in any capacity for a number of years fairly entitles the person who has rendered it to support in sickness and old age. The demand on the funds of an institution to afford this would never be heavy, and the expenditure would be a fixed one, and incapable of any abuse.

This latter alone would be needed, if the nursing were entrusted to members of any association. The former, however, becomes most desirable as a means of promoting those habits of thrift in which the English as compared with the French are so deficient, if the nurses are lay women not committed to a life work as if they were members of a community or were bound by any engagement, implied or expressed, to more than the honest fulfilment of the charge they undertake, for so long as they choose to continue it.

The number of women who, if untrammelled by any engagement, will choose to devote the whole of their lives to nursing, will most likely be comparatively small. For the majority, nothing can be more desirable than that after some years of faithful service they should become the wives of honest men in their own station of life. For themselves and for their influence for good in the world, this would be far happier and better than that they should be taught to look upon their work as something so exceptional in character, so above the level of ordinary women's duties, as to call for more than average endowments or more than average religious principle to carry it out ; still less such as should imply on their part a desire to renounce the ordinary life of women of their class for its proper performance.

It would require the advice and help of those best acquainted with the organisation of savings' banks and provident institutions to settle the details of the project. Probably something of, this sort might be found not unworkable. Each increase of wages should consist partly of money paid to the nurse, partly of money paid to the fund, which fund should be increased either by a certain proportion added from the funds of the hospital, or by the voluntary contributions of individuals. Every nurse on leaving the hospital should receive the amount of her deposit, with interest thereon at five per cent., which she should either receive in money or have transferred to her account at some savings' bank.

If Englishwomen can once be induced to feel the same pride in bringing some dowry to their husband as the Frenchwoman does in her *Dot*, or in its equivalent, there would be fewer unhappy households than at present among our working classes.

It would be useless to endeavour to follow through every detail the organisation of a hospital. The patients' DIETARY, however, is of necessity one of the chief sources of expense; and, like English food in general, it is too unvaried, and, in the case of children at least, too heavy. Hence arise the frequent deviations from the diet-table, the ordering of so-called fancy diets, to the despair of the housekeeper, and the extreme increase in the difficulty of keeping her accounts.

It is very important therefore for the convenient working of the economical arrangements of a hospital that there should be a minimum of extras, and that the diet-table should be drawn up so that a certain number of diets—1, 2, 3, 4, as the case may be— shall represent to a very great extent the daily requirements of the patients. In the French hospitals [1] there are three chief classes of diets; patients on simple diet, "*aux bouillons;*" patients on broth diet, "*aux potages;*" patients on meat diet, "*aux alimens solides;*" of which latter there are four different scales. The introduction into a hospital dietary of a class of diets the cost of which must vary exceedingly and the demand for it likewise, as a fish diet, for instance, is a grave economical error, since the elements of such a diet are too fluctuating in price, and the demand for it is too uncertain for it to be made the subject of contracts with any purveyor. Whatever too is determined on as the diet-table of any hospital, or whatever modifications are made in a diet scale already existing, everything in either case should at first be merely tentative. The knowledge —to a great extent theoretical—of the medical staff should not be allowed at once to settle questions of so much importance to the hospital expenditure; of which most naturally they as a body neither take nor can be expected to take heed. It may, for instance, be the opinion of one of the staff that a change

---

[1] See the *Règlement sur le Régime Alimentaire des Hopitaux et Hospices Civils de Paris.* 4to. Paris, 1872.

in a particular diet would be advantageous. His colleagues agree to it, but before it is permanently adopted it should be referred to the housekeeper to carry out the change for a month, or for some longer time if needed, to test it, and then to report on the difference of expense between the old diet and its modification. The medical men would have the opportunity of reporting at the end of the same time on the result of the change as far as the condition of the patients was concerned. The equation would not be a difficult one to work,—given, so much additional cost, for every change involves expense, and so much advantage to the patient,—is the latter sufficient to justify the former? In almost all English hospital diet-tables, too, there is an inexactness of details with reference to the preparation of the food, which contrasts most disadvantageously with the minute rules laid down [1] in the Paris hospitals with reference to the exact weight of salt, pepper, herbs, vegetables, &c., to be added to the "*potage*," and the ingredients to be employed in the composition of the different dishes. It would be well worth the while of the managers of any English hospital to obtain for three or six months the help of some sister who had had the superintendence of the kitchen of one of the large Paris hospitals ; and whose suggestions, allowing for all the difference between French and English habits, would almost certainly have the twofold result of rendering the diet both more varied and less costly.

THE WASHING, which forms in all hospitals a large item of expense, is often swelled in cost by needless extravagance, such as the use of good sheets for draw-

1 See, for instance, *Règlement, &c.*, p. 48, art. 89.

sheets; the change of the top sheet for the sake of appearance only; the use of white counterpanes, the undertaking the washing for the patients (though in a children's hospital there is some justification for this for the sake of preventing the importation of contagious fever), and the want of strictness as to the quantity of washing allowed to the nurses and servants.

The expense is also vastly increased wherever no laundry exists in connection with the hospital, and still more when the washing, instead of being given by contract to one large establishment, is divided among two or three small ones. But whether there is a laundry or not connected with the hospital, the disposal of the foul linen is attended by no small difficulty, especially in those hospitals in which each ward is not provided with its own foul linen shoot. In this respect the hospital Lariboisière at Paris may serve as a model. Each ward in that hospital has its own foul linen shoot, which runs down external to the building into a large box or closet which opens on the outside, but is kept shut under lock and key except at the time for emptying it every morning, when it is at once conveyed to the foul linen room and counted, and then sent to the laundry in the grounds of the hospital. At the Children's Hospital the narrow space on which the building stands did not leave room for such an arrangement: and there is but one foul linen shoot common to each of three wards, which are situated one above the other on either side of the building. There is an opening, closed by an iron door, which communicates with the shoot on each landing of the back or service stairs.

The following plan was adopted there, and was

found to work well. In the linen room adjoining each ward a change of everything is kept, sufficient to supply the demand for each bed for 48 hours; and this is far preferable to the plan of reserving any large store in the linen room, since it ensures the more equal wear of all articles, and much facilitates the keeping the inventory of each ward.

When the linen is changed, every evening and morning, the foul linen is put into a bag marked with the name of the ward, and thrown down the shoot; the number of each article being marked by one of the nurses, under the supervision of the sister, in a book to which is attached a counterfoil, as in the copy annexed.

HELENA WARD.				HELENA WARD.			
	187	TOTAL.			E.	M.	TOTAL.
Supply				*Receive*			
Children's Sheets	.	.		Children's Sheets .			
Draw ditto.	.	.		Draw ditto .	.		
Pillow Cases	.	.		Pillow Cases .	.		
Towels	.	.		Towels .	.		
Roller ditto	.	.		Roller ditto .	.		
Bath ditto .	.	.		Bath ditto .	.		
Counterpanes	.	.		Counterpanes	.		
Table Cloths	.	.		Table Cloths .			
Dusters	.	.		Dusters .	.		
Sundries	.	.		Sundries	.		
Blankets, Upper	.	.		Blankets, Upper .			
Do. Under	.	.		Do. Under .			
Night Gowns	.	.		Night Gowns	.		
Do. Jackets .	.	.		Do. Jackets	.		
Day ditto .	.	.		Day ditto	.		
Tea Cloths .	.	.		Tea Cloths	.		
Glass ditto .	.	.		Glass ditto	.		

————————————Sister.   ——————————————Sister.

Signature of }————————   Signature of }————————
Head of Linenry }   Head of Linenry }

Any linen which may require to be changed in the interval between the two countings is placed in a bin

G

lined with zinc with a tight-fitting lid; and a perforated false bottom, beneath which is a drawer filled with carbolic acid powder. The contents of this box are included in each counting, and the total of the evening and morning countings is added up every morning in the third column and signed by the sister of the ward.

Every morning a nurse from each ward attends in the foul linen room, and counts out the contents of her bags before the head of the linenry, who affixes her signature in verification of the accuracy of the check, which she tears off and files. She likewise signs the counterfoil, which should correspond with the numbers on the check, and places that on a separate file. The numbers on the checks furnish the material for making out the washing list, those on the counterfoil the quantity of clean articles needed; and which should correspond exactly with those of the dirty linen left.

The foul linen room is divided into bunkers corresponding with the nature of the different articles enumerated in the washing list; which plan greatly facilitates the sorting of the foul linen.

Any articles which are very dirty are placed in a tank containing a mixture of carbolic acid and water: and when thus freed from any offensive matters are passed through one of Bradford's wringers, by which they are dried sufficiently to allow of their being loosely folded, and thrown each into its own bunker. It further diminishes trouble if in sending to the wash all articles are put up in tens, twenties and fifties.

In the afternoon a nurse goes from each ward to claim her supply of clean linen, which is put ready for her in the linenry, taking with her a requisition for so-called medical stores, as lint, oiled silk, &c.; such requisition being previously signed by the sister of the ward, and kept by the head of the linenry, who is answerable to the housekeeper for the management of her department.

The above arrangement implies that all linen is sent to the wash every second day, and nothing can be more detrimental to the linen or more injurious to the health of those employed in the hospital than keeping foul linen stored for days, before sorting and sending it to the wash.

Some apology is needed perhaps for these somewhat dreary details. The idea of the plan was derived in a measure from the Paris hospitals. It seems to work very well, and some modification of it may be found specially useful in small hospitals unfurnished with laundries, and even without foul linen shoots.

No review, however hasty, of hospital organisation can be regarded as complete in which no reference is made to the OUT-PATIENT DEPARTMENT.

It probably very rarely happens that the in-patient department of a hospital is abused by a person seeking admission whose resources would enable him to command medical attendance at home. It may indeed be said that the opposite tendency prevails to a degree very injurious to the chances of recovery from serious illness of persons whose resources are small. The English working class shrink from the inside of the hospital very often as if it were the ante-chamber of the workhouse; and persuasion is often needed to induce any of them to enter it, or to send to it members of their own family, rather than to urge them to cultivate the spirit of independence, and not needlessly to seek refuge within its walls.

With the out-patient department, however, the case is altogether different. The impression, and that not altogether a mistaken one, that the advice obtained at a hospital is that of more skilled or more specially experienced persons than those who are the ordinary medical attendants of himself and family; the temptation, too, to get what he believes to be a superior article without paying for it, are to many almost irresistible. The belief too that these institutions are wealthy, and that their officers are highly paid, takes away that sense of honour which would make the working man shrink from appropriating to himself the food, or the clothing, or the alms destined for the poor. The temptation, too, to avoid doctors' bills, or to get for nothing the opinion of some eminent man to whom it might not be quite convenient to pay his guinea, have affected a class far above that of the well-to-do artizan, and have thus led to an intolerable abuse of the out-patient department of our hospitals. It is not only those who would be able by ordinary thrift to obtain attendance for themselves and their families by a small weekly payment when in health to a provident dispensary, who abuse a charity intended for the really poor; but shopkeepers, and others of the middle class, are not ashamed to put on shabby clothes and to present themselves as candidates for that succour on which they have no claim.

Hence has arisen not merely an enormous waste of public money, nor only, in addition, a most unfair demand upon the time of the medical men attached to the different institutions, but the whole system of out-patient attendance has been reduced to an absurdity, by the numbers of applicants exceeding all human power to attend to properly. When patients are seen, as they sometimes are, at the rate of nearly sixty in an hour, the highest aim of the doctor must needs be limited to the endeavour to do no harm. Of doing good it is clear that there can be no question.

Hence there have arisen of late years from various quarters loud reclamations concerning the abuses of the out-patient departments of hospitals ; and hence the well-founded complaints of the medical officers, that their function was rendered a mockery by the afflux of people in numbers so great, that it was impossible to pay them proper attention. It has been attempted to lessen these evils by the inculcation of habits of thrift among the artizan class, and by the establishment, as far as possible, of provident dispensaries.[1] The authorities of hospitals, too, have in a measure co-operated by endeavouring to lessen the number of applicants. But too often they have done this in the most unintelligent of all ways : that of fixing an arbitrary limit to their number, without any attempt at discriminating between deserving and non-deserving cases, while at the same time they made an exception in favour of persons being the bearers of letters of recommendation from governors; though such letters are given away, almost invariably with so little thought that they furnish absolutely no guarantee for the fitness of those who present them.

[1] One of the good works of the Charity Organisation Society, and one in which Sir C. Trevelyan has taken a prominent part, has been the promotion of the establishment of provident dispensaries. A full account of their working recently will be found in Sir C. Trevelyan's pamphlet, *Metropolitan Medical Relief*. 8vo. London, 1877.

In these circumstances, the first step towards laying down rules for the prevention of the abuse of the out-patient department of a hospital is to determine what objects it should really attempt to fulfil.

1st. It should provide for the immediate relief of cases of accident and emergency.

2nd. It should afford the advice of specially skilled persons in cases of doubt or difficulty.

3rd. It should provide for the continued treatment of cases, which for some reason or other are not received, or may even not be admissible, into the hospital, but which require for their cure constant supervision, or appliances, or remedies beyond the reach of all but the comparatively wealthy.

4th. It should supply both advice and medicine in ordinary ailments to persons, who though above receiving parochial relief, are yet unable to meet the cost of an extra such as medical assistance.

There can be no question as to the importance of the first three of these objects, or as to the desirability of restricting as far as may be the work of the out-patient department of a hospital to their fulfilment ; and to making its medical staff a sort of court of appeal from all the provident dispensaries, or other medical institutions in the neighbourhood, in cases involving doubt or difficulty.

It will however, most probably, never be possible to eliminate entirely, at any rate in a large city, the claims to consideration of the fourth class. In London, for instance, there must be many thousands of the population suddenly reduced from a position of something more than comfort to one bordering on penury, and many who are but temporary residents in the metropolis ; or whose business compels them to change their dwellings frequently from one part of the city to another. If to these are added the improvident or thoughtless, who have trusted that good health would always be their portion, and have needlessly neglected when well to provide for the possibility of sickness, there will be a very large number of persons, to some of whom it would be wrong, and to others, to say the least very harsh, to refuse all help when overtaken by illness, and they will always make a fourth class too numerous and too important altogether to ignore.

The question then is how to provide for them in such a way as to prevent the recurrence of those abuses which it is so important to avoid.

The attempt was made at the Children's Hospital to provide for the accomplishment of the first three objects, and at the same time to prevent the abuses so difficult to check in carrying out the fourth. The machinery adopted for this purpose is perhaps better explained by reprinting the regulations than in any other way.

REGULATIONS.

" That all who apply as out-patients shall be sorted by the house-surgeon in attendance into two classes, viz. : those of slighter ailments, to be treated as casualties, *i.e.* to be seen and prescribed for once, and graver ailments, to be entered on the register, and to be supplied with letters, entitling the bearer to attend for a period of two months.

" The register and also the letter shall bear a record of the name and residence of the patient, as also of the occupation and weekly earnings of the parent on whom the patient is dependent ; and no patient whose parent is in the receipt of more than 30s. per week shall be considered eligible for permanent treatment as an out-patient of the hospital.

" No patient shall be considered eligible for *permanent* relief as an out-patient unless the letter with which the parent has been furnished shall have been submitted to the secretary or other officer of the Charity Organisation Society for the district in which the patient dwells, and unless such letter bears the stamp of the society,[1] in proof of the verification of the facts stated in the letter.

" No patient whose parents are in receipt of parish relief shall be considered eligible for *permanent* treatment as an out-patient.

" With reference to this class of applicants, however, as also with reference to those whose parents are in the receipt of more than 30s. per week, it shall be in the power of the medical officers to dispense with these regulations *on purely medical grounds*, provided always that such medical officer enter in a book, to be kept for that purpose, the particulars concerning such patient, and the reason for the exemption, and sign it with his own name. The said book to be laid before the managing committee at each meeting.

" That no letter be valid for more than *two* months, unless the medical officer sees fit to prolong the time for the patient's attendance, in which case he shall supply the patient with a new letter, indicating that the case is chronic. Such chronic patient may be brought at such intervals of time as the medical officer may desire ; but if medicine is prescribed, such

---

[1] Each letter bore the address of the agent of the Charity Organisation Society for every district of the metropolis.

medicine shall in no case whatever be supplied for a longer period than one week, though it may be obtained weekly, by application at the dispensary, without its being prescribed afresh.

"That in order to prevent patients continuing their attendance beyond the period of two months, without the authority of the medical officer, the dispenser be instructed to stamp each letter, the period of which is about to expire, so that the attention of the medical officer may be called to the case on the occasion of the patient's next visit."

It will be observed that on the first attendance every applicant was admitted without any question being asked as to who he was or whence he came. It happened, too, occasionally that mothers came stating that they did not desire to attend permanently, but that they wished for an opinion concerning their children in some respect or other. These were in many instances persons who confessed that they did not stand in need of the constant help of a hospital, but that they had some doubt they wished solved, or anxiety which they sought to have removed.

Further, to provide against the exclusion on social grounds of any of the third class of patients, the medical officers were empowered on *purely medical* grounds to except any case from the rule, either because of its medical interest, or because the treatment it required would be of necessity either costly or protracted.

Every case therefore being seen at first and judged of by the physician or surgeon, there remained only the fourth class of applicants, namely, those seeking for permanent relief in ordinary ailments, with whom the aim was to deal as wisely and, at the same time, as gently as possible.

The question of the husband's earnings was almost always answered, approximately at least, by the wife: and the agent of the Society had only to verify her statement, which he could do in many instances, even without a visit to her house, from his own personal knowledge. This system worked so efficiently that the number of out-patients was reduced from 13,000 in 1874 to 9,000 in 1876.

No change can be introduced in any system, however bad, without its being distasteful to some persons; and the out-patient department of the hospital has now relapsed into its former mode of working, controlled only by a selection of a limited number of cases for permanent treatment by the house-surgeons, that is to say, by the youngest and least experienced officers of the institution.

The merits and demerits of this or of any other plan can be weighed only by a consideration of the objections which have been regarded as sufficient to warrant its discontinuance.

The first reclamation against it, and one which was accepted as valid, came from the young gentlemen, the house-surgeons, who alleged that it was beneath the dignity of the medical profession to have to put questions to the patients concerning their husbands' earnings or concerning any matter not purely medical.

Next came the complaint of subscribers, who considered that a guinea subscription on their part should give a qualification to every bearer of a letter of recommendation, overriding all inquiry; and some in consequence withdrew their contributions. According to this view, while it is open to the doctor to say on medical grounds that a case is unfit for hospital relief, the stewards of the public money, who administer the affairs of the institution, are to be precluded from determining, with the approval of the governors, for what social class the hospital is intended. It is more than probable that a few words of explanation from the committee as to the reasons for the regulation would have satisfied all but a very small minority of the subscribers.

Lastly, complaints were made in a few exceptional instances of the inquisitorial nature of the questions put by the agents of the Charity Organisation Society, though it was universally allowed that on the whole their work had been done with both wisdom and gentleness, and the spirit in which they were instructed to act was that of endeavouring to admit all applicants save those manifestly unfit. In many cases, indeed, these agents stated to the secretary exceptional circumstances with reference to many of the applicants which justified the admission for permanent treatment of persons whom the strict letter of the law would have excluded.

The bearing of many regulations is not immediately apparent, and the essential difference between the position of the officers of the Charity Organisation Society when called on merely to verify a statement, or when required to institute an entirely new investigation, does not perhaps lie on the surface. The patients scarcely ever demurred to answering the inquiry as to their husbands' earnings when put to them at the hospital, where they were sure of receiving, once at any rate, the advice they sought, and invariably a few words of kindly explanation got over all difficulty. The case, however, was widely altered when they were referred elsewhere, and called on to reply to questions which it had not been thought necessary to propose at the hospital, while the agent of the society had no data furnished to guide his inquiries. He at once was placed in the offensive position of the Relieving Officer of the Parish, who is supposed popularly to derive his name from its being his duty to refuse relief. The failure of the plan was secured by this apparently insignificant change, just as surely as the upset of a carriage by the removal of the linch-pin.[1]

[1] The out-patient difficulty is not confined to England, or to London, but is felt as much in France and Paris ; and the evil of unchecked admission was found in 1874 to have attained such dimensions at Lyons that the municipal council of that city adopted a regulation requiring that no person should be admitted to the *consultations* at the hospitals unless furnished with a recent certificate of indigence from the Bureau de Bienfaisance. The inquiries made by the officer of the Bureau are of the most extreme minuteness, and this whether the patient is to be visited at his own home, or merely to be entitled to receive medicine gratuitously as an out-patient. The person even who seeks merely to have local treatment for his children for ringworm at the Hopital St. Louis, the great hospital for diseases of the skin in Paris, is forced to make a statement, among other things, as to his occupation, his daily wage, and the amount of his rent. These statements are made by the writer with all the different out-patient papers of the Paris hospitals before him, but copies of them will be found in a little book of regulations issued by the administration of hospitals, and still in force. *Instruction pour l'exécution de l'Arrêté, du 20 Avril, 1853, portant organisation du service du traitement des malades à Domicile.* 8vo. Paris, 1853.

The new stringency introduced by the municipality of Lyons into the regulations for out-patients led to much discussion, and the reader who cares to follow it, and it is by no means without profit to do so, will find both sides of the question handled in *L'Union Médicale* for 1874, Nos. 84, 86, 99, 100, and 103. The chief practical suggestion which was the outcome of the debate was that those who sought for gratuitous medical relief at the hospitals should provide themselves with a special card from the Bureau, stating that they were not indigent but *nécessiteux*. This plan too is in work in the second arrondissement of Paris ; and the name of Dr. Boinet, who advocates its general adoption, will carry weight with all members of the medical profession. See *L'Union Médicale*, No. 99, August 18, 1874, p. 256. This plan is exactly the same as it was sought to carry out by the help of the Charity Organisation Society, of distinguishing between the pauper and those who were yet unable to meet the exceptional demand of illness on their resources.

The difficulties in the way of the judicious administration of medical relief are much the same in all countries. They who care to pursue the inquiry, especially with reference to medical relief, will find ample

The inquiry if made only at the hospital, as has been proposed by some, would not answer the purpose, for a hospital, and particularly a special hospital in a large city like London, attracts to its out-patient room people from distant parts, concerning whom all local knowledge would be impossible, and any check on the abuse of a hospital and any other charity must be real as well as apparent.

It may be added that before calling in the aid of the Charity Organisation Society the attempt had been made to prevent the hospital out-patient room being resorted to by those who had no real claim upon its help, by requiring that before presenting themselves a second time each patient's letter should be counter-signed by some householder, clergyman, city missionary, or other minister of religion. The regulation turned out a farce, the clergy of no persuasion cared to be guilty of the apparent unkindness of refusing their signature to a person asking for it merely to obtain hospital relief, and the small shopkeeper in a poor district did not dare to disoblige his customers. Theoretically this test of fitness seemed to conciliate kindness and discrimination, practically it was worthless.

It has already been pointed out that the establishment of provident dispensaries, how widely soever those institutions may be disseminated, cannot en-

material for the study in Davenne, *op. cit.*, vol. ii. chap. v. ; *Règlement Administratif sur les Secours à Domicile dans la Ville de Paris.* 8vo. Paris, 1871. *Compte Moral de l'Administration de l'Assistance Publique.* 4to. Paris, 1874, § vi. Gyoux, *Du Service Médical des Pauvres.* 8vo. Versailles, 1868. Neboux, *Projet d'Organisation de l'Assistance Publique, dans la ville de Paris, &c.* 8vo. Paris, 1850. Rause, *Réorganisation de l'Assistance Publique.* 8vo. Paris, 1871. Pétrequin, *De l'Organisation de l'Assistance Publique à Lyon.* 8vo. Lyons, and the valuable work of M. de Gerando, vol. iv. livre ii., already referred to, besides many others.

tirely supersede the calls upon a hospital out-patienr room even in cases of ordinary illness. On the othe hand, it will not be found possible in a city such as London to affiliate the provident dispensary as a rule to any of the larger and more important hospitals. A provident dispensary is essentially local in its cha- racter; one of its great objects is to cultivate thrift among the working classes; its administration must be carried on in large measure in the way in which that of a club or benefit society is conducted; and the committee who manage it must be composed chiefly of those who join it for the sake of themselves pro- fiting by it, and in only a small proportion of those who are honorary members in virtue of their voluntary contributions to its funds; or of the influence which they exert to promote its success. The machinery of a large hospital is too complicated to admit of the placing in intimate connection with it a number of smaller institutions, each of which would encounter something of the same class of difficulties as the parent itself has to meet with, and is, like it, exposed to all the risks attendant on ignorance, incompetence, and want of business habits on the part of those who undertake its management.

When, lastly, it has been proposed to prevent the abuse of hospitals by admitting persons to the out- patient room on payment of one shilling the first time and threepence or other small sum on each subsequent visit, the great principle that should be borne in mind in the administration of all forms of charity—that of interfering as little as possible with the social life of the poor, and with the business or occupation of those whose livelihood depends on supplying their daily wants—is altogether lost sight of.

Such a plan ministers in no way whatever to the promotion of thrift; for it is entirely antagonistic to the important object of providing against the evil day by the exercise of foresight during the season of health and prosperity. It would at once lead to a grievous abuse of the charity by the comparatively rich ; and it would ruin the general practitioner in the poorer districts of the metropolis—the man who attends the wife in her confinement, and the husband in his sickness, and who cannot live out of mere casual emergencies if he has to meet the competition of the hospitals. This man, however, forms a most important element in the social life of the poor, of whom he is often the friend and counsellor, no less than is the clergyman, or dissenting minister, or priest, or city missionary. The suggestion is altogether as great a breach of the commonest principles of political economy as it would be to establish with the aid of public subscriptions a society for the supply of the poor with meat, or vegetables, or groceries at a lower rate than that at which they were furnished by the butcher or the chandler, or the costermonger in their own neighbourhood.

It may be hoped, then, that this plan will be given up, and that, while promoting as far as possible the establishment of provident dispensaries, some such mode as has already been suggested and tried with success may be had recourse to to check the abuse of the out-patient department of our hospitals.

Here end these suggestions on hospital organisation, fragmentary as from their nature they must be, though they have touched on all the main principles which should underlie the management of institutions for the relief of the sick poor. It is hoped that they do not contain one word that could seem to express a harsh judgment of any who, though with the mingled motives which influence us all, even in our noblest undertakings, devote their time and their personal service to the work of charity.

But the same laws as those on whose observance depends the success of any commercial enterprise govern also that of any charitable undertaking, and determine especially the prosperity of any hospital. The only difference is this, that the dividends in the one case are paid here, in the other they are received elsewhere, and the coin they are paid in bears the stamp of no earthly mint. The medical report and the treasurer's balance-sheet furnish the test of the good working of a hospital. If the mortality is excessive, if the expenditure is extravagant, there is something wrong, something which must be set right—considerately always, gently if possible—but which yet must be set right, cost what it may.

We profess it to be our aim to tread in the footsteps of Him who was Himself the embodiment of the highest charity. We shall surely then do well to follow His example, when He whose resources were as boundless as His compassion bade His disciples, after He had fed the five thousand with five barley loaves and two small fishes, "to gather up the fragments, that nothing be lost."

And now in laying down my pen I may once again speak in the first person, which I have carefully avoided through these pages. I believe the facts stated to be correct, but mistakes often creep in one knows not how, and suggestions made can doubtless be improved by the added experience of others.

The foundation of a Hospital for Sick Children was

the dream of my youth, and the occupation of thirty years of manhood. I looked forward to helping to organise the institution which is housed in a building planned by me in conjunction with the architect, Mr. E. Barry, and which, like the old hospital, has been fitted and furnished throughout under my direction ; and I trusted that when old age came I might be allowed to linger about the place, towards which my heart turns as a parent's towards his child.

But, *Dîs aliter visum est*, and there is nothing left for me but to commend this little book to the serious consideration of those who have undertaken to carry on my work. Counsel sometimes has more weight when the personality of the counsellor is no longer obtruded on those whom he ventures to advise.

THE END.

www.ingramcontent.com/pod-product-compliance
Lightning Source LLC
Chambersburg PA
CBHW021943190326
41519CB00009B/1128